New Horizons in Vision Science, Optometry and Ocular Surface

New Horizons in Vision Science, Optometry and Ocular Surface

Editors

José-María Sánchez-González
Carlos Rocha-de-Lossada
Alejandro Cerviño

MDPI

Basel • Beijing • Wuhan • Barcelona • Belgrade • Novi Sad • Cluj • Manchester

Editors

José-María Sánchez-González
Department of Physics of Condensed Matter, Optics Area
University of Seville
Seville
Spain

Carlos Rocha-de-Lossada
Department of Surgery
University of Seville
Seville
Spain

Alejandro Cerviño
Department of Optics & Optometry & Vision Sciences
University of Valencia
Valencia
Spain

Editorial Office
MDPI
St. Alban-Anlage 66
4052 Basel, Switzerland

This is a reprint of articles from the Special Issue published online in the open access journal *Life* (ISSN 2075-1729) (available at: www.mdpi.com/journal/life/special_issues/89ZL3SBLQF).

For citation purposes, cite each article independently as indicated on the article page online and as indicated below:

Lastname, A.A.; Lastname, B.B. Article Title. *Journal Name* **Year**, *Volume Number*, Page Range.

ISBN 978-3-0365-9065-3 (Hbk)
ISBN 978-3-0365-9064-6 (PDF)
doi.org/10.3390/books978-3-0365-9064-6

Contents

life

MDPI

Editorial

Exploring the Cutting Edge of Vision Science: New Developments in Diagnostics and Treatment of Ocular Surface in Dry Eye Disease

José-María Sánchez-González [1,*], Carlos Rocha-de-Lossada [2,3,4,5] and Alejandro Cerviño [6]

1. Department of Physics of Condensed Matter, Optics Area, University of Seville, 41012 Seville, Spain
2. Qvision, Ophthalmology Department, VITHAS Almeria Hospital, 04120 Almeria, Spain
3. Ophthalmology Department, VITHAS Malaga, 29016 Malaga, Spain
4. Regional University Hospital of Malaga, Hospital Civil Square, 29009 Malaga, Spain
5. Surgery Department, Ophthalmology Area, University of Seville, Doctor Fedriani, 41009 Seville, Spain
6. Department of Optics & Optometry & Vision Sciences, University of Valencia, 46010 Valencia, Spain

* Correspondence: jsanchez80@us.es

check for updates

Citation: Sánchez-González, J.-M.; Rocha-de-Lossada, C.; Cerviño, A. Exploring the Cutting Edge of Vision Science: New Developments in Diagnostics and Treatment of Ocular Surface in Dry Eye Disease. *Life* **2023**, *13*, 1584. https://doi.org/10.3390/life13071584

Received: 7 July 2023
Accepted: 17 July 2023
Published: 19 July 2023

The ocular surface refers to the outermost layer of the eye, which includes the cornea, conjunctiva and eyelids [1]. It is a complex and delicate system that is responsible for maintaining the health and function of the eye [2]. One common problem that can affect the ocular surface is dry eye syndrome, which is a multifactorial condition that occurs when the eye does not produce enough tears or the tears produced do not have the correct balance of water, mucus and oil, where inflammation, hyperosmolarity and neurosensorial abnormalities could coexist [3]. Dry eye syndrome can be caused by a variety of factors, including aging, certain medications and environmental conditions, such as prolonged use of the screen or living in a dry or dusty environment [3], or be related to autoimmune syndromes, where Sjögren syndrome stands out [4]. Symptoms and signs of dry eye syndrome include burning, itching or redness of the eye, as well as a sensation of dryness, grittiness or a foreign body sensation, among others. In severe cases, dry eye syndrome can cause vision problems, corneal damage and even blindness [5,6]. Effective treatment of dry eye syndrome requires an accurate diagnosis and a personalized treatment plan. Currently, the most common subjective method of diagnosing dry eye syndrome is through the use of questionnaires and subjective tests, such as the Ocular Surface Disease Index (OSDI) [7] or the Dry Eye Questionnaire (DEQ) [7]. Although these tests can be useful in identifying the presence of dry eye syndrome, they are based on self-reported symptoms and may not accurately reflect the true severity of the condition [7]. Moreover, traditional objective tests include the Oxford grading system with fluorescein stain or lissamine green stain, the tear break-up time test (TBUT) or the Schirmer test (I and II) [1–10]. Other adjuvant options are the measurement of the tear meniscus [11], the evaluation of the metalloproteinases [12] or the tear osmolarity [13].

A new measurement device, called the Ocular Surface Analyzer (OSA) [14], has recently been developed to help improve the diagnosis and management of dry eye syndrome. The OSA is a non-invasive, objective tool that uses interferometry to measure the thickness and surface profile of the tear film, similar to a Keratograph [15]. It can also measure blink rate and eyelid position, as well as TBUT. One of the key advantages of the OSA is its ability to provide detailed, quantitative data about the ocular surface [14]. This allows for a more accurate diagnosis of dry eye syndrome and allows for a more personalized treatment plan to be developed [16]. The OSA can also be used to monitor the effectiveness of treatment and track changes in the ocular surface over time [17]. There is growing evidence to suggest that OSA is a valuable tool in the diagnosis and management of dry eye syndrome [18]. A recent study found that OSA was able to accurately detect changes in the ocular surface in patients with dry eye syndrome and that it was able to distinguish between different severities of the condition [19]. Another study found that OSA was able to accurately predict the presence of dry eye syndrome in patients who had previously been diagnosed with the condition [20].

In general, OSA appears to be a promising new tool for the diagnosis and treatment of dry eye syndrome. Its non-invasive, objective measurements provide detailed, quantitative data about the ocular surface, which can help to improve the accuracy of diagnosis and the effectiveness of treatment. As research on OSA and its potential uses continues to be conducted, it is likely to become an increasingly valuable tool in the treatment of dry eye syndrome and other conditions that affect the ocular surface.

Second, dry eye disease is a common condition that occurs when the eye does not produce enough tears or the tears evaporate too quickly [21]. This can lead to a variety of symptoms, including dryness, irritation, redness and a feeling of discomfort or a foreign body sensation [7]. In severe cases, it can even cause vision problems [5]. The stability of the tear film is an important factor in the health and comfort of the eye [22–24]. The tear film is a thin layer of moisture that coats the surface of the eye and helps to keep it lubricated and protected [25]. It is made up of three layers: the outer layer, which is composed of oil produced by the meibomian glands [26,27]; the middle layer, which is made up of water produced by the lacrimal glands; and the inner layer, which is composed of mucus produced by the conjunctiva [10].

The tear film plays a critical role in maintaining the health of the eye by providing a protective barrier against dust, dirt and other irritants. It also helps keep the eye surface moist and comfortable, which is essential for good vision [6]. There are several factors that can affect tear film stability, including age, hormonal changes [16,28,29], medications [30,31] and certain medical conditions. Dry eye disease is one of the most common causes of tear film instability and is more common in women than in men, especially after menopause [16,28,29]. To treat dry eye disease, doctors may recommend the use of eye drops or ointments to supplement the natural tear film [6,9]. There are several types of eyedrops available, including artificial tears, which are designed to mimic the natural tear film; lubricating drops, which help to moisturize the eye; and anti-inflammatory drops, which reduce inflammation and redness [28–33].

In recent years, there has been a significant amount of research into the development of new eyedrop formulations that are more effective in treating dry eye disease. One promising area of research is the use of lipid-based eyedrops, which are designed to mimic the natural tear film more closely [33]. These drops are composed of a mixture of oils and water, and they are able to stay on the eye longer than traditional artificial tears. Another area of research is the use of nanotechnology to create eyedrops with smaller particle sizes that can be absorbed more easily by the eye [32]. These drops have the potential to provide more sustained relief from dry eye symptoms and may be more effective in improving tear film stability [22]. In general, dry eye disease is a common and often debilitating condition that can cause a variety of symptoms, including dryness, irritation and vision problems. Maintaining the stability of the tear film is an important factor in maintaining eye health and comfort, and new eyedrop formulations are being developed to improve the treatment of dry eye disease [34–40]. In more severe cases, other options, such as the use of corticosteroids, autologous or allogenic serum [41], immunomodulators or secretagogues and even surgical approaches [42], are necessary [6].

On the last point of discussion for this Editorial, the ocular surface microbiota, or the collection of microorganisms living on the surface of the eye, has long been recognized as an important factor in the health and function of the eye [43]. However, the specific role of the ocular surface microbiota in the development and management of eye diseases is still not fully understood [44]. Recently, a multicenter study proposed the concept of eye community state type (ECST) as a way to categorize and understand the different profiles of bacterial communities that can exist in the healthy eye [43]. The study found that nine different ECSTs could be identified within the healthy bacterial population. This is an exciting finding, as it suggests that there may be multiple "healthy" states of the ocular surface microbiota and that different individuals may have different ECSTs. It also opens up the possibility of developing personalized approaches to eye care based on an individual's ECST.

However, more research is needed to fully understand the clinical implications of ECST and how it may be related to the development and management of eye diseases [45–48]. For example, it is not yet known whether certain ECSTs are more or less prone to developing eye infections or other problems. Overall, the concept of ECST is an interesting new avenue for research on the ocular surface microbiota and its role in eye health [49–51]. Further studies are needed to fully understand the clinical importance of ECST and how it can be used to improve the diagnosis and treatment of eye diseases. The field of vision science is constantly evolving, and there have been many exciting developments in the diagnosis and treatment of the ocular surface in dry eye disease. New technologies and approaches are being developed that have the potential to greatly improve the lives of those suffering from this common and often debilitating condition.

An important area of research is the development of new eyedrop formulations that are more effective at treating dry eye disease. Lipid-based eyedrops, which mimic the natural tear film, and nanotechnology-based drops with smaller particle sizes, which can be absorbed more easily by the eye, are both promising approaches that have the potential to provide more sustained relief from dry eye symptoms and improve tear film stability. In addition, advances in diagnostic techniques, such as the use of non-invasive imaging techniques, are helping to improve the accuracy and reliability of dry eye diagnoses. This is important because it allows physicians to more effectively tailor treatment plans to the specific needs of each patient.

In general, the cutting edge of vision science provides new and innovative ways to diagnose and treat ocular surface conditions, including dry eye disease. These developments have the potential to greatly improve the lives of those affected by this condition and to help them maintain the health and comfort of their eyes.

Conflicts of Interest: The authors declare no conflict of interest.

References

1. Nelson, J.D.; Craig, J.P.; Akpek, E.K.; Azar, D.T.; Belmonte, C.; Bron, A.J.; Clayton, J.A.; Dogru, M.; Dua, H.S.; Foulks, G.N.; et al. TFOS DEWS II Introduction. *Ocul. Surf.* **2017**, *15*, 269–275. [CrossRef] [PubMed]
2. Craig, J.P.; Nichols, K.K.; Akpek, E.K.; Caffery, B.; Dua, H.S.; Joo, C.-K.; Liu, Z.; Nelson, J.D.; Nichols, J.J.; Tsubota, K.; et al. TFOS DEWS II Definition and Classification Report. *Ocul. Surf.* **2017**, *15*, 276–283. [CrossRef] [PubMed]
3. Stapleton, F.; Alves, M.; Bunya, V.Y.; Jalbert, I.; Lekhanont, K.; Malet, F.; Na, K.-S.; Schaumberg, D.; Uchino, M.; Vehof, J.; et al. TFOS DEWS II Epidemiology Report. *Ocul. Surf.* **2017**, *15*, 334–365. [CrossRef]
4. Barrientos, R.T.; Godín, F.; Rocha-De-Lossada, C.; Soifer, M.; Sánchez-González, J.-M.; Moreno-Toral, E.; González, A.-L.; Zein, M.; Larco, P.; Mercado, C.; et al. Ophthalmological Approach for the Diagnosis of Dry Eye Disease in Patients with Sjögren's Syndrome. *Life* **2022**, *12*, 1899. [CrossRef] [PubMed]
5. Bron, A.J.; de Paiva, C.S.; Chauhan, S.K.; Bonini, S.; Gabison, E.E.; Jain, S.; Knop, E.; Markoulli, M.; Ogawa, Y.; Perez, V.; et al. TFOS DEWS II pathophysiology report. *Ocul. Surf.* **2017**, *15*, 438–510. [CrossRef] [PubMed]
6. Jones, L.; Downie, L.E.; Korb, D.; Benitez-Del-Castillo, J.M.; Dana, R.; Deng, S.X.; Dong, P.N.; Geerling, G.; Hida, R.Y.; Liu, Y.; et al. TFOS DEWS II Management and Therapy Report. *Ocul. Surf.* **2017**, *15*, 575–628. [CrossRef]
7. Wolffsohn, J.S.; Arita, R.; Chalmers, R.; Djalilian, A.; Dogru, M.; Dumbleton, K.; Gupta, P.K.; Karpecki, P.; Lazreg, S.; Pult, H.; et al. TFOS DEWS II Diagnostic Methodology report. *Ocul. Surf.* **2017**, *15*, 539–574. [CrossRef] [PubMed]
8. Belmonte, C.; Nichols, J.J.; Cox, S.M.; Brock, J.A.; Begley, C.G.; Bereiter, D.A.; Dartt, D.A.; Galor, A.; Hamrah, P.; Ivanusic, J.J.; et al. TFOS DEWS II pain and sensation report. *Ocul. Surf.* **2017**, *15*, 404–437. [CrossRef]
9. Craig, J.P.; Nelson, J.D.; Azar, D.T.; Belmonte, C.; Bron, A.J.; Chauhan, S.K.; de Paiva, C.S.; Gomes, J.A.P.; Hammitt, K.M.; Jones, L.; et al. TFOS DEWS II Report Executive Summary. *Ocul. Surf.* **2017**, *15*, 802–812. [CrossRef]
10. Willcox, M.D.P.; Argüeso, P.; Georgiev, G.A.; Holopainen, J.M.; Laurie, G.W.; Millar, T.J.; Papas, E.B.; Rolland, J.P.; Schmidt, T.A.; Stahl, U.; et al. TFOS DEWS II Tear Film Report. *Ocul. Surf.* **2017**, *15*, 366–403. [CrossRef]
11. Rocha-de-Lossada, C.; Sánchez-González, J.M.; Zamorano-Martín, F.; Rachwani-Anil, R.; Torras-Sanvicens, J.; Peraza-Nieves, J. Influence of Sodium Hyaluronate Concentration in Tear Meniscus Height: 10-min Dynamic Profile After Single Instillation. *Eye Contact Lens* **2021**, *47*, 330–334. [CrossRef] [PubMed]
12. Huh, J.; Choi, S.Y.; Eom, Y.; Kim, H.M.; Song, J.S. Changes in the Matrix Metalloproteinase 9 Point-of-Care Test Positivity According to MMP-9 Concentration and Loading Volume. *Cornea* **2020**, *39*, 234. [CrossRef]

13. Lemp, M.A.; Bron, A.J.; Baudouin, C.; Bentez Del Castillo, J.M.; Geffen, D.; Tauber, J.; Foulks, G.N.; Pepose, J.S.; Sullivan, B.D. Tear osmolarity in the diagnosis and management of dry eye disease. *Am. J. Ophthalmol.* **2011**, *151*, 792–798.e1. [CrossRef] [PubMed]
14. Sánchez-González, M.C.; Capote-Puente, R.; García-Romera, M.-C.; De-Hita-Cantalejo, C.; Bautista-Llamas, M.-J.; Silva-Viguera, C.; Sánchez-González, J.-M. Dry eye disease and tear film assessment through a novel non-invasive ocular surface analyzer: The OSA protocol. *Front. Med.* **2022**, *9*, 938484. [CrossRef] [PubMed]
15. García-Marqués, J.V.; Martínez-Albert, N.; Talens-Estarelles, C.; García-Lázaro, S.; Cerviño, A. Repeatability of Non-invasive Keratograph Break-Up Time measurements obtained using Oculus Keratograph 5M. *Int. Ophthalmol.* **2021**, *41*, 2473–2483. [CrossRef] [PubMed]
16. García-Marqués, J.V.; Talens-Estarelles, C.; García-Lázaro, S.; Wolffsohn, J.S.; Cerviño, A. Systemic, environmental and lifestyle risk factors for dry eye disease in a mediterranean caucasian population. *Contact Lens Anterior Eye* **2022**, *45*, 101539. [CrossRef]
17. Singh, S.; Srivastav, S.; Modiwala, Z.; Ali, M.H.; Basu, S. Repeatability, reproducibility and agreement between three different diagnostic imaging platforms for tear film evaluation of normal and dry eye disease. *Eye* **2022**, *45*, 101539. [CrossRef] [PubMed]
18. Lee, J.M.; Jeon, Y.J.; Kim, K.Y.; Hwang, K.-Y.; Kwon, Y.-A.; Koh, K. Ocular surface analysis: A comparison between the LipiView® II and IDRA®. *Eur. J. Ophthalmol.* **2021**, *31*, 2300–2306. [CrossRef]
19. Rinert, J.; Branger, G.; Bachmann, L.M.; Pfaeffli, O.; Iselin, K.; Kaufmann, C.; Thiel, M.A.; Baenninger, P.B. Accuracy of a New Noninvasive Automatic Ocular Surface Analyzer for the Diagnosis of Dry Eye Disease-Two-Gate Design Using Healthy Controls. *Cornea* **2022**, *42*, 416–422. [CrossRef]
20. Yadav, S.; Gupta, N.; Makwana, T.; Vanathi, M.; Tandon, R. Noninvasive ocular surface analyzer as an adjunct in diagnosis and estimating prevalence of meibomian gland dysfunction: Hospital-based comparative study. *Indian J. Ophthalmol.* **2022**, *70*, 1539–1545. [CrossRef]
21. Montés-Micó, R.; Cerviño, A.; Ferrer-Blasco, T.; García-Lázaro, S.; Madrid-Costa, D. The tear film and the optical quality of the eye. *Ocul. Surf.* **2010**, *8*, 185–192. [CrossRef] [PubMed]
22. Montés-Micó, R.; Cerviño, A.; Ferrer-Blasco, T.; García-Lázaro, S.; Ortí-Navarro, S. Optical quality after instillation of eyedrops in dry-eye syndrome. *J. Cataract Refract. Surg.* **2010**, *36*, 935–940. [CrossRef]
23. García-Marqués, J.V.; Macedo-De-Araújo, R.J.; McAlinden, C.; Faria-Ribeiro, M.; Cerviño, A.; González-Méijome, J.M. Short-term tear film stability, optical quality and visual performance in two dual-focus contact lenses for myopia control with different optical designs. *Ophthalmic Physiol. Opt. J. Br. Coll. Ophthalmic Opt.* **2022**, *42*, 1062–1073. [CrossRef] [PubMed]
24. Capote-Puente, R.; Eftimov, P.; Bautista-Llamas, M.-J.; Yokoi, N.; Sánchez-González, J.-M.; Georgiev, G. Short-term tear film stability, optical quality and visual performance in two dual-focus contact lenses for myopia control with different optical designs. *Ophthalmic Physiol. Opt. J. Br. Coll. Ophthalmic Opt.* **2022**, *43*, 290–291. [CrossRef] [PubMed]
25. Vidal-Rohr, M.; Wolffsohn, J.S.; Davies, L.N.; Cerviño, A. Effect of contact lens surface properties on comfort, tear stability and ocular physiology. *Cont. Lens Anterior Eye* **2018**, *41*, 117–121. [CrossRef]
26. García-Marqués, J.V.; García-Lázaro, S.; Martínez-Albert, N.; Cerviño, A. Meibomian glands visibility assessment through a new quantitative method. *Graefe's Arch. Clin. Exp. Ophthalmol.* **2021**, *259*, 1323–1331. [CrossRef]
27. García-Marqués, J.V.; Talens-Estarelles, C.; García-Lázaro, S.; Cerviño, A. Validation of a new objective method to assess lipid layer thickness without the need of an interferometer. *Graefe's Arch. Clin. Exp. Ophthalmol.* **2022**, *260*, 655–676. [CrossRef]
28. De-Hita-Cantalejo, C.; Sánchez-González, M.C.; Silva-Viguera, C.; García-Romera, M.C.; Feria-Mantero, R.; Sánchez-González, J.-M. Efficacy of hyaluronic acid 0.3%, cyanocobalamin, electrolytes, and P-Plus in menopause patients with moderate dry eye disease. *Graefe's Arch. Clin. Exp. Ophthalmol.* **2022**, *260*, 529–535. [CrossRef]
29. Serrano-Morales, J.-M.; De-Hita-Cantalejo, C.; Sánchez-González, M.C.; Bautista-Llamas, M.-J.; Sánchez-González, J.-M. Efficacy of 0.1% crosslinked hyaluronic acid, coenzyme Q10 and vitamin E in the management of dry eye disease in menopause patients receiving antidepressants. *Eur. J. Ophthalmol.* **2022**, *32*, 658–663. [CrossRef]
30. Sánchez-González, M.C.; De-Hita-Cantalejo, C.; Martínez-Lara, C.; Sánchez-González, J.-M. Oral isotretinoin for acne vulgaris side effects on the ocular surface: Hyaluronic acid and galacto-xyloglucan as treatment for dry eye disease signs and symptoms. *Front. Med.* **2022**, *9*, 959165. [CrossRef]
31. Sánchez-González, J.-M.; De-Hita-Cantalejo, C.; Sánchez-González, M.C. Hyaluronic Acid and Galacto-Xyloglucan Eyedrop Efficacy in Young-Adult Oral Contraceptive Users of Childbearing Age. *J. Clin. Med.* **2022**, *11*, 4458. [CrossRef] [PubMed]
32. Sánchez-González, J.-M.; De-Hita-Cantalejo, C.; Sánchez-González, M.C. Crosslinked hyaluronic acid with liposomes and crocin for management symptoms of dry eye disease caused by moderate meibomian gland dysfunction. *Int. J. Ophthalmol.* **2020**, *13*, 1368–1373. [CrossRef] [PubMed]
33. Sánchez-González, J.-M.; De-Hita-Cantalejo, C.; Martínez-Lara, C.; Sánchez-González, M.C. Lipid, Aqueous and Mucin Tear Film Layer Stability and Permanence within 0.15% Liposome Crosslinked Hyaluronic Acid versus 0.15% Non-Crosslinked Hyaluronic Acid Measured with a Novel Non-Invasive Ocular Surface Analyzer. *J. Clin. Med.* **2022**, *11*, 3719. [CrossRef] [PubMed]
34. Talens-Estarelles, C.; García-Marqués, J.V.; Cerviño, A.; García-Lázaro, S. Determining the Best Management Strategy for Preventing Short-Term Effects of Digital Display Use on Dry Eyes. *Eye Contact Lens* **2022**, *48*, 416–423. [CrossRef]
35. Szczesna-Iskander, D.H.; Muzyka-Wozniak, M.; Llorens Quintana, C. The efficacy of ocular surface assessment approaches in evaluating dry eye treatment with artificial tears. *Sci. Rep.* **2022**, *12*, 21835. [CrossRef] [PubMed]
36. Chen, N.; Zhang, J.-S.; Zhang, T.-X.; Fan, B.-L.; Ning, Y. The effect of sodium hyaluronate on tear film stability in patients with dry eye syndrome after cataract surgery. *Graefe's Arch. Clin. Exp. Ophthalmol.* **2022**, *261*, 1011–1017. [CrossRef]

37. Srinivasan, S.; Williams, R. Propylene Glycol and Hydroxypropyl Guar Nanoemulsion—Safe and Effective Lubricant Eye Drops in the Management of Dry Eye Disease. *Clin. Ophthalmol.* **2022**, *16*, 3311–3326. [CrossRef]
38. Roszkowska, A.M.; Spinella, R.; Oliverio, G.W.; Postorino, E.I.; Signorino, G.A.; Rusciano, D.; Aragona, P. Effects of the Topical Use of the Natural Antioxidant Alpha-Lipoic Acid on the Ocular Surface of Diabetic Patients with Dry Eye Symptoms. *Front. Biosci. Landmark Ed.* **2022**, *27*, 202. [CrossRef]
39. Murtaza, F.; Toameh, D.; Chiu, H.H.; Tam, E.S.; Somani, S. Autologous Platelet-Rich Plasma Drops for Evaporative Dry Eye Disease from Meibomian Gland Dysfunction: A Pilot Study. *Clin. Ophthalmol.* **2022**, *16*, 2199–2208. [CrossRef]
40. Montani, G.; Landini, L.; Martino, M. Short- and Long-Term Effects of a Multi-Component, Artificial Tear on Preocular Tear Film Stability, Tear Evaporation and Tear Film Optical Dynamic: A Prospective Randomized Double-Phase Study. *Curr. Eye Res.* **2022**, *47*, 1252–1258. [CrossRef]
41. Rodríguez Calvo-de-Mora, M.; Domínguez-Ruiz, C.; Barrero-Sojo, F.; Rodríguez-Moreno, G.; Antúnez Rodríguez, C.; Ponce Verdugo, L.; Hernández Lamas, M.d.C.; Hernández-Guijarro, L.; Villalvilla Castillo, J.; Fernández-Baca Casares, I.; et al. Autologous versus allogeneic versus umbilical cord sera for the treatment of severe dry eye disease: A double-blind randomized clinical trial. *Acta Ophthalmol.* **2022**, *100*, e396–e408. [CrossRef] [PubMed]
42. Messmer, E.M.; Ahmad, S.; Benitez Del Castillo, J.M.; Mrukwa-Kominek, E.; Rolando, M.; Vitovska, O.; Baudouin, C.; A Panel of European Dry Eye Disease Experts. Management of inflammation in dry eye disease: Recommendations from a European panel of experts. *Eur. J. Ophthalmol.* **2022**, *33*, 1294–1307. [CrossRef]
43. Borroni, D.; Paytuví-Gallart, A.; Sanseverino, W.; Gómez-Huertas, C.; Bonci, P.; Romano, V.; Giannaccare, G.; Rechichi, M.; Meduri, A.; Oliverio, G.W.; et al. Exploring the Healthy Eye Microbiota Niche in a Multicenter Study. *Int. J. Mol. Sci.* **2022**, *23*, 10229. [CrossRef] [PubMed]
44. Gallon, P.; Parekh, M.; Ferrari, S.; Fasolo, A.; Ponzin, D.; Borroni, D. Metagenomics in ophthalmology: Hypothesis or real prospective? *Biotechnol. Rep.* **2019**, *23*, e00355. [CrossRef] [PubMed]
45. Borroni, D.; Romano, V.; Kaye, S.B.; Somerville, T.; Napoli, L.; Fasolo, A.; Gallon, P.; Ponzin, D.; Esposito, A.; Ferrari, S. Metagenomics in ophthalmology: Current findings and future prospectives. *BMJ Open Ophthalmol.* **2019**, *4*, e000248. [CrossRef] [PubMed]
46. Hernández-Zulueta, J.; Navarro-Partida, J.; Sánchez-Aguilar, O.E.; Cruz-Pavlovich, H.D.S.; Castro-Castañeda, C.R.; la Rosa, A.G.-D. An insight on the eye bacterial microbiota and its role on dry eye disease. *APMIS* **2022**, *131*, 103–111. [CrossRef] [PubMed]
47. Tong, L.; Constancias, F.; Hou, A.; Chua, S.L.; Drautz-Moses, D.I.; Schuster, S.C.; Yang, L.; Williams, R.B.H.; Kjelleberg, S. Shotgun metagenomic sequencing analysis of ocular surface microbiome in Singapore residents with mild dry eye. *Front. Med.* **2022**, *9*, 1034131. [CrossRef]
48. Watane, A.; Raolji, S.; Cavuoto, K.; Galor, A. Microbiome and immune-mediated dry eye: A review. *BMJ Open Ophthalmol.* **2022**, *7*, e000956. [CrossRef]
49. An, Q.; Zou, H. Ocular surface microbiota dysbiosis contributes to the high prevalence of dry eye disease in diabetic patients. *Crit. Rev. Microbiol.* **2022**, 1–10. [CrossRef]
50. Jing, D.; Jiang, X.; Ren, X.; Su, J.; Huang, C.; Yang, J.; Hao, R.; Li, X. Metagenomic nanopore sequencing of ocular microbiome in patients with meibomian gland dysfunction. *Front. Med.* **2022**, *9*, 1045990. [CrossRef]
51. Chen, Z.; Jia, Y.; Xiao, Y.; Lin, Q.; Qian, Y.; Xiang, Z.; Cui, L.; Qin, X.; Chen, S.; Yang, C.; et al. Microbiological Characteristics of Ocular Surface Associated with Dry Eye in Children and Adolescents with Diabetes Mellitus. *Investig. Ophthalmol. Vis. Sci.* **2022**, *63*, 20. [CrossRef] [PubMed]

life

Article

Visual Quality, Motility Behavior, and Retinal Changes Associated with Reading Tasks Performed on Electronic Devices

Elvira Orduna-Hospital *, María Munarriz-Escribano and Ana Sanchez-Cano *

Department of Applied Physics, University of Zaragoza, 50009 Zaragoza, Spain; mariamunarriz00@gmail.com
* Correspondence: eordunahospital@unizar.es (E.O.-H.); anaisa@unizar.es (A.S.-C.)

Abstract: Background: The purpose of this study was to objectively evaluate visual discomfort using an eye tracker and aberrometer after a 21-min reading session on an iPad and an Ebook. Additionally, retinal changes were analyzed using optical coherence tomography (OCT). Methods: A total of 31 young subjects (24 ± 4 years) participated in this study. They read for 21 min on an Ebook and for another 21 min on an iPad under controlled lighting conditions while their eye movements were monitored using an eye tracker. Aberrometry and retinal OCT measurements were taken before and after each reading session. Parameters such as pupil diameter, fixations, saccades, blinks, total aberration, high-order aberration, low-order aberration, and central and peripheral retinal thickness in the nine early treatment diabetic retinopathy study (ETDRS) areas were measured for each reading situation. Statistical analysis was performed on the collected data. Results: No statistically significant differences ($p > 0.05$) between the two devices were observed in terms of the different types of eye movements or the changes in retinal thickness. However, the aberrometric analysis showed variations in post-reading situations depending on the device used. Conclusion: Reading speed and visual discomfort resulting from electronic device usage can be objectively assessed using an eye tracker and aberrometer. Additionally, changes found in central and peripheral retinal thickness between the two devices and the baseline measurements were not significant and remained relatively stable.

Keywords: eye tracker; eye movements; aberrometry; visual quality; optical coherence tomography

check for updates

Citation: Orduna-Hospital, E.; Munarriz-Escribano, M.; Sanchez-Cano, A. Visual Quality, Motility Behavior, and Retinal Changes Associated with Reading Tasks Performed on Electronic Devices. *Life* **2023**, *13*, 1777. https://doi.org/10.3390/life13081777

Academic Editors: Nicola Smania, José-María Sánchez-González, Alejandro Cerviño and Carlos Rocha-de-Lossada

Received: 4 July 2023
Revised: 12 August 2023
Accepted: 18 August 2023
Published: 20 August 2023

1. Introduction

Recently, the use of near vision (NV) has been steadily increasing for the execution of various tasks, whether they involve work, study, or entertainment. Activities such as reading, performing tasks, and viewing videos or movies on digital platforms are increasingly being carried out using electronic devices (computers, tablets, or smartphones).

The consequences of the continuous and prolonged use of these devices result in various ocular symptoms, such as irritation, dryness, itching, tearing, burning, or a foreign body sensation [1,2], leading to visual fatigue, diplopia, blurred vision, or asthenopia, among other conditions. These signs mainly appear in individuals who spend four or more consecutive hours using these devices, with symptoms intensifying with daily use exceeding seven hours [3,4], affecting the attention and performance of individuals who experience them; this condition is known as computer vision syndrome (CVS) [5]. It is estimated that the prevalence of CVS can range from 25% to 93%, depending on the device used, the duration of use, and environmental factors. The number of people affected is projected to increase significantly each year [3,4].

Objective signs also appear, such as changes in accommodation and optical manifestations such as ocular aberrations in response to prolonged NV work [6,7]. Wick and Morse demonstrated that the accommodative response (accommodative lag) increases in computer work [8]. However, Moulakaki et al. argued that the accommodative response is independent of the device used [9]. Other studies show that both low-order aberrations (LOAs) and high-order aberrations (HOAs) increase with accommodation, particularly spherical aberrations. These changes may be due to alterations in the different ocular media during the accommodation process, resulting in an imperfect image formed on the retina [10]. Changes in retinal thickness and shape have also been observed during reading under different lighting conditions, especially in low light conditions in multiple macular areas [11].

CVS is a prevalent condition that can cause discomfort and affect productivity and quality of life; there is no universally agreed-upon definition or diagnostic criteria, and the best interventions are not well established [2]. It is known that computer use reduces the frequency and amplitude of blinking; however, for handheld devices such as tablets, e-books, mobile phones, etc., this has not been conclusively proven, although it is assumed that they affect tear stability. Despite this finding, there are no irrefutable studies to confirm that discomfort or other symptoms such as asthenopia caused by device use is related to blinking [12].

There are two main types of e-readers: e-books, which are designed to resemble printed books on paper, and liquid crystal display (LCD) screens, which are used in most electronic devices such as iPads or tablets. Reading on an LCD display can cause more visual fatigue than reading a paper book or E-ink display, likely due to the higher level of luminance emitted by the LCD display, which can cause the pupil to constrict and the frequency of eye blinks to decrease. These changes can lead to eye strain and fatigue [13]. In spite of this, one advantage of using these technologies is the ability to customize factors such as font type and size, screen luminance, and even background color for reading, allowing for individualized adaptation to the user's preferences [13].

Reading speed is a measure of how quickly a person can read and comprehend text. It is an important indicator of reading fluency, which is essential for academic success, as it allows students to read and understand complex texts in a timely manner. In general, reading speed increases with age and may be affected by visual discomfort, which is caused by several factors, including reading for long periods of time, reading in poor lighting conditions, or reading a text that is too small or too difficult [14].

Some studies have focused on how ambient lighting, contrast, and the difference in luminance between the background and text affect visual comfort when reading on electronic devices [15]. Others have specifically investigated the optimal ambient lighting conditions. It is specified that the minimum illumination should be 200 lux for comfortable and pleasant reading [16]. If the ambient lighting is too high, the visibility of the screens may be compromised, contrast can be lost, and reading can become more challenging. However, for E-ink devices without backlighting, the ambient lighting needs to be even higher, preferably above 700 lux, to ensure effective readability [17].

The reading process involves saccades, which are small and quick jumps lasting approximately 20–40 ms with which the gaze direction is changed, causing the image of the object of interest to remain on the fovea. These are binocular movements, where both eyes (OU) move in the same direction, performing conjugated movements (versions) following the line of the text being read [18,19]. Following a saccade, a fixation occurs, which is the time in which the gaze remains on the object of interest for approximately 200 to 250 ms; these are the moments in which reading is carried out [19]. Fixations do not occur word by word but rather involve reading groups of words, valued as reading efficiency [20]. The concept of "regression" refers to fixations that occur from right to left, backward movements in reading to reread a word or group of words, and movement to the next line [21].

Generally, proficient readers employ a lower number of fixations and spend less time on them compared to less skilled readers. The latter group experiences more difficulty comprehending the text as they analyze the meaning of each word rather than taking a global approach to the context [21]. Many individuals struggle to comprehend what they read or find that they require a significant amount of time to read and understand, leading to frustration and inhibiting skill development. This highlights the importance of emphasizing reading fluency and comprehension at an early age [22].

The eye tracker is an electronic device that records gaze tracking, allowing for the measurement of unconscious eye movements during specific tasks. It provides insights into the skills and cognitive processes of the subject being evaluated. Eye tracking is a scientific research method used in various fields, including advertising [23], psychology [24], human–computer interaction [25], and some optometric tests [26–28]. The gaze-mind hypothesis is investigated by examining individuals' preferences based on the direction of their visual axes [29].

The primary objective of this study was to assess reading speed, monitored with an eye tracker, to evaluate visual discomfort, aberrometric changes, and retinal thickness in different quadrants after sustained NV tasks using an iPad or e-book under controlled lighting conditions.

2. Materials and Methods

2.1. Sample Description and Selection

The study was conducted in accordance with the principles outlined in the Declaration of Helsinki and with the approval of the Clinical Research Ethics Committee of Aragón (CEICA) under reference number PI21-074 and with signed informed consent from the participants. The sample consisted of 31 healthy individuals, 20 females and 11 males, with an age range between 18 and 31 years. A comprehensive optometric evaluation was performed on the participants, including measurement of the best-corrected visual acuity (BCVA) in both distance vision (DV) and NV, monocular accommodative amplitude, accommodative and convergence facility in both monocular and binocular vision, associated and dissociated phoria measurement, positive and negative fusional vergences at near and distance, and fusion and stereopsis; ocular motility was also assessed. Participants who required optical correction were asked to bring their contact lenses since the antireflective coating is designed for the wavelength range of 400–700 nm; it also reflects other wavelengths, including the infrared used by the eye tracker (>750 nm). Subjects who met the exclusion criteria were not able to participate in the experiment. These criteria included binocular vision problems, BCVA less than 0.8 decimal in one of the eyes, vision-impairing pathologies, media opacities, dry eye syndrome, use of electronic devices within one hour before the measurements, consumption of coffee, smoking, engaging in high-intensity exercise, or attending the session without their contact lenses corrected for DV since they were young participants and had enough accommodation to focus on NV.

2.2. Devices Used, Setup, and Lighting

One of the instruments used for the measurements in the study was the Tobii Pro Fusion Eye Tracker (Tobii AB, Danderyd, Sweden). This device operates by emitting infrared light at around 850 nm that produces a corneal reflection captured by a camera within the device. By analyzing the corneal reflection and pupil position, the device determines the direction of the visual axes and estimates where the subject is looking. Prior to reading, individual calibration is performed for each person by directing their gaze to calibration points displayed on the screen to be used, providing information about the duration of saccades and fixations [30].

Two different electronic devices were used for reading: an 8th generation iPad, Model A2270 (Apple Inc., Cupertino, CA, USA), with screen dimensions of 250.6 × 174 × 7.5 mm and 2160 × 1620 pixels and an E-ink reader Ebook (ink pad 3, PocketBook International, Lugano, Switzerland), model PB740, with dimensions of 195 × 136.5 × 8 mm and 1872 × 1404 pixels. Times New Roman font was used with a size of 9 pixels for the iPad and 10 pixels for the Ebook (slightly different by the resolution described for each screen). Thus, visual acuity was achieved of around 0.8 decimal in both devices when reading at 50 cm.

During the reading sessions, recordings were made, requiring a camera with a microphone connected to the laptop from which the readings were monitored using eye tracker software. The exact camera model used was AMDIS01B (Conceptronic, Dortmund,

Germany), while the software programs employed were the Eye tracker Manager (Tobii AB, Danderyd, Sweden) for selecting the device used for reading and the Tobii Pro Lab (Tobii AB, Danderyd, Sweden) for individual calibration of each subject for each reading session.

The experimental components were placed within a light control cabinet. In addition, two additional tools were used for the measurements: a chin rest and a stand to hold the reading device (Figure 1).

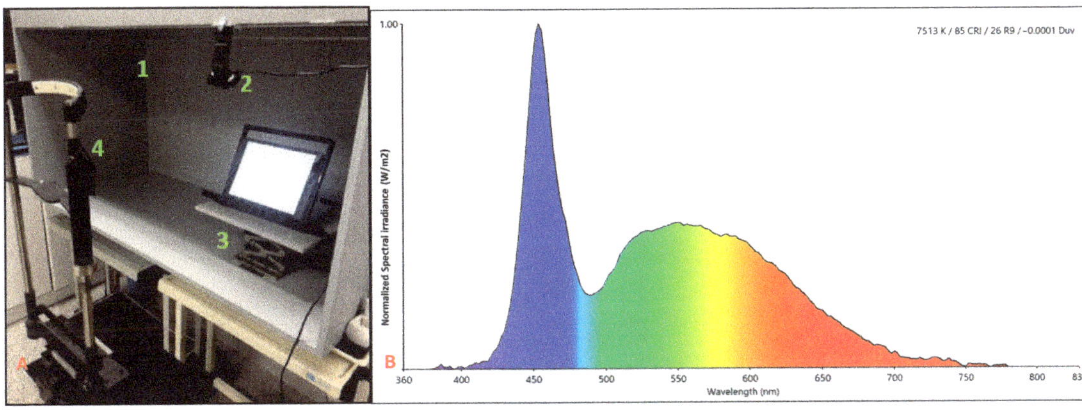

Figure 1. Image (**A**) Study elements. (1) Light control cabinet. (2) Camera. (3) Stand for device and eye tracker support. (4) Chin rest. Image (**B**) Normalized spectral irradiance (W/m^2) of the ambient light reaching the corneal plane while reading.

The experiment consisted of two readings: one using an iPad and the second using an Ebook, randomly assigned, with both readings conducted under controlled lighting conditions. To determine the irradiance (W/m^2) and illuminance (lux) on the corneal plane and the luminance (cd/m^2) of the reading devices, a spectroradiometer (model StellarNet-Black Comet, StellarNet, Inc., Tampa, FL, USA with C20080502 calibration and NIST traceability) and a luminance meter (Mavo-Spot 2, Gossen-Kainos, Barcelona, Spain) were used. The luminance perceived by the eye and emitted by the iPad was 59.57 cd/m^2 and 58.01 cd/m^2 for the Ebook. The illuminance reaching the corneal plane was 257.0 lux for both the iPad and the Ebook and the irradiance was 0.91 (W/m^2) for the iPad and 0.87 (W/m^2) for the Ebook, as shown in Figure 1.

The analysis of visual quality was performed using the IRX3 Hartmann–Shack aberrometer (Image Eyes, Orsay, France). This device uses a light source of 780 nm that is projected onto the retina and, based on the impact of the rays coming from it on a CCD, generates a map of the total aberration of the evaluated eye.

The study of retinal changes was carried out by capturing images of multiple retinal layers using the 3D OCT-1000 model (Topcon Corporation, Tokyo, Japan). The light source was a superluminescent diode with a wavelength of 840 nm and a bandwidth of 50 nm. The longitudinal (depth) resolution was 6 μm (A-scan) and the maximum transverse (horizontal) resolution was 20 μm (B-scan).

The protocol followed with each participant involved obtaining baseline measurements of OU separately using the aberrometer and four measurements using OCT, two for each eye, in a random order each time, always under scotopic lighting conditions. The participant was asked to focus on the central square that appeared, thus capturing the image of the central 30° of the retina centered on the fovea. The macular cube protocol was performed, capturing 128 tomographic retinal slices, with the macula in the center of the image (Figure 2A,B). The second captured image involved looking toward the temporal end of the central line, thus obtaining 128 baseline images of the temporal peripheral retina using the same macular cube protocol but with the eye rotated 15° toward the temporal side (Figure 2C,D).

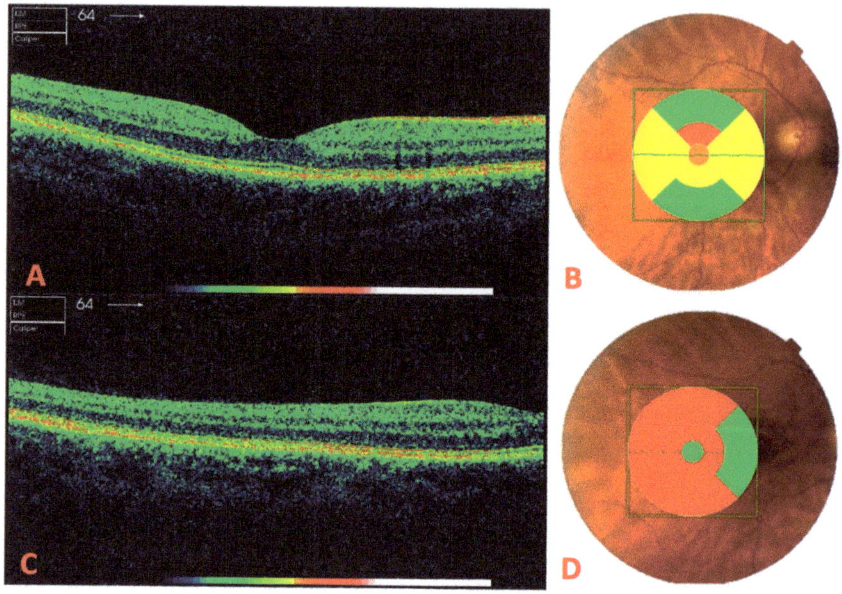

Figure 2. Image (**A**) A tomographic cross-section of the central retina. Image (**B**) The fundus of the eye with central retinal thickness in the 9 quadrants of the early treatment diabetic retinopathy study (ETDRS). Image (**C**) A tomographic cross-section of the peripheral retina. Image (**D**) The fundus of the eye with peripheral retinal thickness in the 9 quadrants of the ETDRS. All these images correspond to the right eye (OD) of the same individual.

The proposed two readings were performed 50 cm from the reading device using either the Ebook or the iPad. The eye tracker was calibrated for each subject and each reading device, ensuring that the eye tracker detected OU. A calibration template consisting of numbers from 1 to 5 was used. The subject was instructed to sequentially fixate on each number until all of them had been completed. From the computer, the examiner could observe which point the subject was looking at and accept or reject the calibration.

The reading session lasted for 21 min. The examiner recorded one minute of reading every 5 min to evaluate the reading process and assess the eye movements and visual discomfort throughout the reading session (Figure 3). This resulted in a total of five-1 min recordings.

Immediately after the 21 min reading session, the same baseline tests were repeated with an aberrometer and OCT. The experiments were performed in the morning and the participants were instructed to take a 15 min break, emphasizing the need to maintain the same conditions as 1 hour before the tests. For the second reading session, the same procedure was followed, but this time, the device that was not used in the first reading session was employed.

2.3. Data Export and Statistical Analysis

The data collected with the aberrometer were exported to an Excel database (Microsoft® Office Excel 2011, Microsoft Corporation, Redmond, WA, USA). The recordings taken with the eye tracker were segmented using Tobii Pro Lab software. The "events" option was used to select the start of reading by the subject and, after 60 s, the segment was cut using the same option. Two markers appeared in the recording to delimit the segment (Figure 4). This process resulted in a total of 5 min divided into 5 different 1 min recordings. After performing the same process for all participants, each recording was individually exported in Excel format. For the analysis of the exported data, a program called Etracker Parse Video (University of Zaragoza, Zaragoza, Spain) was created (Figure 5). This program allowed for the selection of the data of interest during the "events," such as the total duration (s), number (n) of blinks, saccades, fixations, pupil diameter of the left eye (oculus sinister, OS) and right eye (oculus dexter, OD) (mm), length (mm), duration (ms), and velocity (m/s) of the saccades for each eye separately, as well as the average duration of fixations (ms).

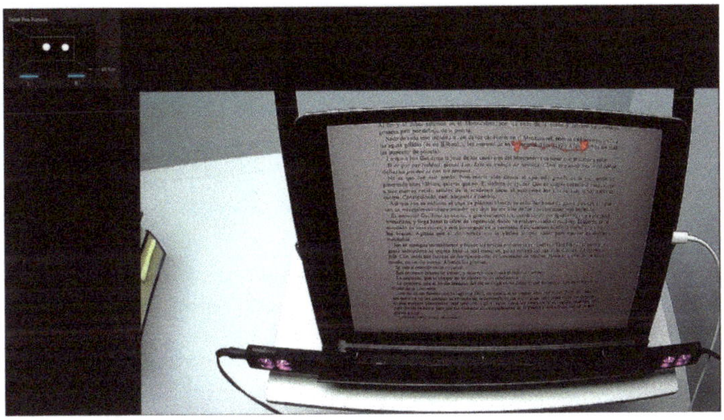

Figure 3. The upper left quadrant indicates the detection of both eyes (OU) of the subject. The red line observed over the text of the reading device (iPad) represents eye movements, tracking the line of text during reading.

Figure 4. Segmentation of the reading recording with the iPad using Tobii Pro Lab. The blue markers indicate the start and end of a 1 min segment of the recording. The red circles indicate the eye movements made during the reading session.

The analysis of all collected data was conducted using the Statistical Package for the Social Sciences (SPSS) version 24.0 (IBM Corp., Armonk, NY, USA). Descriptive statistics were initially performed on the quantitative variables, including the calculation of the mean, standard deviation (±SD), maximum, and minimum values. The normality distribution of all variables was assessed using the Kolmogorov-Smirnov test, which indicated that the sample did not follow a normal distribution. Therefore, nonparametric tests for related samples, specifically the Wilcoxon signed-rank test, were employed to examine differences between the variables when comparing the two reading conditions and, in the case of aberrometry and OCT measurements, comparing them with the baseline measurements. A value of $p < 0.05$ was considered to indicate statistical significance.

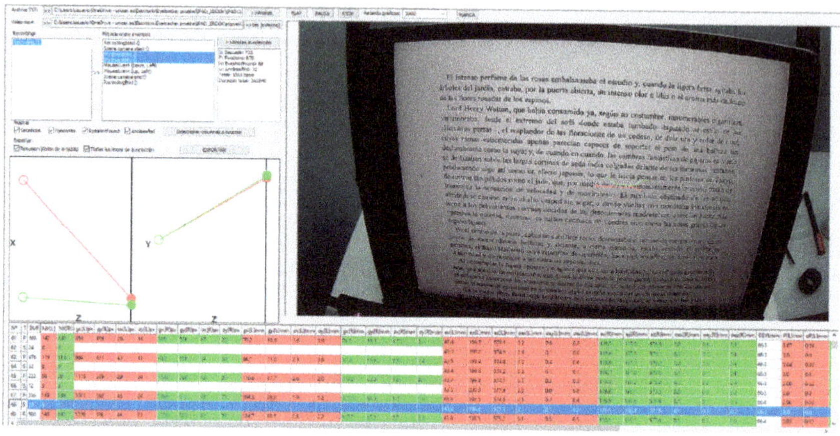

Figure 5. Analysis of a selected event from a reading session with the iPad using the Etracker Parse program. The lines overlaid on the text represent a saccade, which indicates rapid eye movement during the reading process.

3. Results

Thirty-one young subjects (mean age: 24 ± 4 years) who met the inclusion and exclusion criteria described in the previous sections were selected.

3.1. Eye Tracker

Fixations. The number of fixations during reading with the iPad reached its maximum value in the first minute of reading (187.90 ± 23.59), while with the Ebook, it occurred at 10 min (176.45 ± 31.47). In both devices, the minimum number of fixations was observed at the 20th minute, with 176 ± 35.10 and 171.84 ± 32.25 fixations, respectively. However, statistically significant differences were found only at the beginning of the test ($p = 0.002$), as shown in Figure 6A. There were no statistically significant differences in the duration of fixations during the readings (Figure 6B), although, during the first 5 min, these fixations had a longer duration compared to the subsequent time intervals recorded. Overall, it can be observed that the use of the iPad, compared to the Ebook, led to a higher number of fixations on the text during reading, as well as longer fixation durations.

Saccades. During reading with the iPad, the maximum number of saccades occurred at the 10th minute (283.13 ± 173.29), while with the Ebook, it was at the 15th minute (271.39 ± 223.51). The minimum number of saccades was observed in the 5th minute (244.42 ± 132.92) for the iPad and in the first minute (229.03 ± 109.82) for the Ebook. In this case, no statistically significant differences were found in the duration or number of saccades (Figure 7). There were no differences observed throughout the reading; they remained consistent in both readings.

In Figure 8, the length and velocity of saccades are shown separately for each eye (monocular); in general, both measures were higher in readings with the iPad, except for the 10th and 20th minutes with the OS, where velocity were higher with the Ebook. No statistically significant differences were found for velocity. However, for saccade length, significant differences were observed in the OS during the first minute and at the 20th minutes and in the OD at the 10th and 15th minutes, with p values of 0.043, 0.020, 0.019, and 0.024, respectively.

Blinks. The number of blinks during reading with the iPad reached its maximum value at the 20th minute (19.76), while, with the Ebook, it was at the 10th minute (19.43). The minimum number of blinks for both devices occurred in the first minute and was 11.32 and 16.59 blinks, respectively. Significant differences were found only in the first minute of the test ($p = 0.025$) (Figure 9). In general, it can be observed that with Ebook reading compared to iPad reading, subjects blinked more, but, after 20 min of reading, the number of blinks tended to equalize.

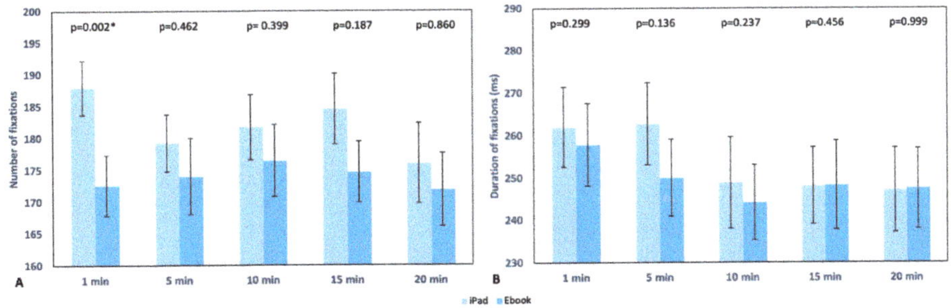

Figure 6. Graphical representation of the number (**A**) and duration (ms) (**B**) of fixations in each reading recording with the iPad and Ebook, respectively. The upper area of the graph displays the *p* value, indicating * statistically significant differences; standard error is also displayed as error bars.

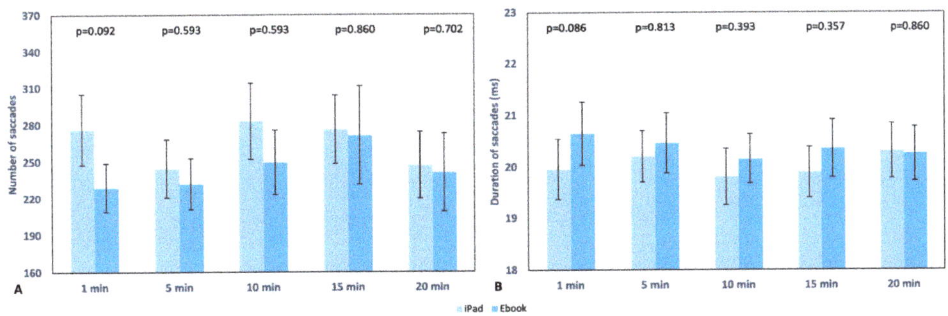

Figure 7. Graphical representation of the number (**A**) and duration (ms) (**B**) of saccades in each reading recording with the iPad and Ebook, respectively. The upper part of the graph displays the *p* value; error bars are also displayed.

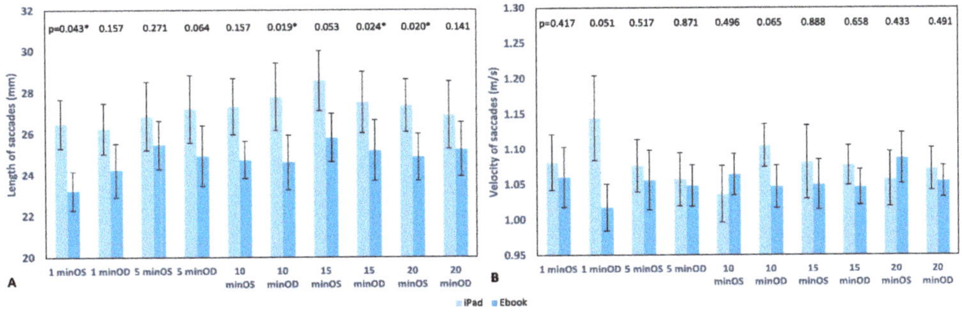

Figure 8. Graphical representation of the length (mm) (**A**) and velocity (m/s) (**B**) of saccades in each reading recording with the iPad and Ebook. The upper part of the graph displays the *p* value, indicating * statistically significant differences; standard error is also displayed as error bars.

Pupil. Throughout all the readings with each device, the pupil diameter remained constant for all subjects, with no significant changes in monocular analysis. When comparing the two reading instruments, significant differences were found at minutes 1 and 5 ($p < 0.05$). The largest recorded mean pupil size was 2.82 ± 0.39 mm in the first minute with the iPad, while the smallest recorded mean pupil size was 2.64 ± 0.31 mm with the Ebook, also during the first minute.

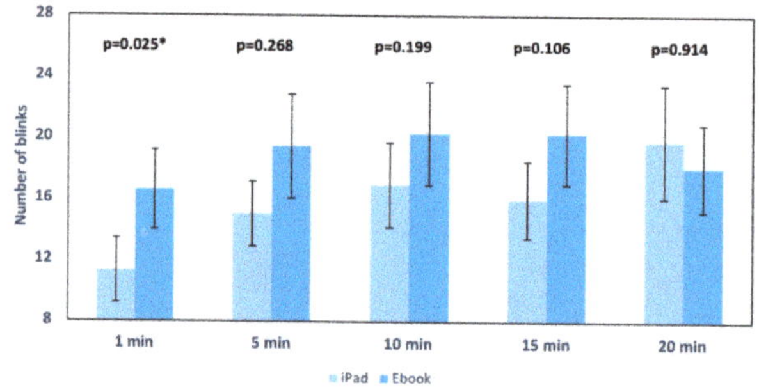

Figure 9. Graphical representation of the number of blinks during each recording in the two performed readings. The upper part of the graph displays the *p* value, indicating * statistically significant differences; standard error is also displayed as error bars.

3.2. *Aberrometry*

In Figure 10A,B, statistically significant differences ($p < 0.05$) can be observed when comparing the total root mean square (RMS_{TOTAL}) between the baseline measurement and each of the reading devices separately, with an increase after reading. The statistical analysis indicated that differences were found in the low-order aberrations root mean square (RMS_{LOA}) ($p = 0.007$ for the Ebook and $p < 0.001$ for the iPad), while the high-order aberrations root mean square (RMS_{HOA}) did not show statistically significant differences in any case. When comparing both devices, no statistically significant differences were found as the *p* value was > 0.05 during all minutes of reading (Figure 10C).

3.3. *Optical Coherence Tomography (OCT)*

Central Retina. Significant retinal thickening ($p = 0.004$) compared to the baseline measurements was found when reading with the Ebook (Figure 11B) in the central ETDRS area. No statistically significant differences were found in Figure 11A,C, when comparing the basal retinal thickness and the retinal thickness after reading with the iPad and with Ebook, respectively. Figure 11G illustrates that there were no statistically significant differences ($p > 0.05$) in macular volume. The highest total volume in the central retina was observed after reading with the iPad, with a value of 7.65 ± 0.38 mm^3, while the minimum volume was 7.62 ± 0.41 mm^3 for the Ebook. In the comparison of average central retinal thickness between baseline measurements and post-reading measurements (Figure 11H), no statistically significant differences were found ($p > 0.05$), with the highest average thickness measured at 270.27 ± 13.54 µm with the iPad and the minimum average thickness at 269.38 ± 14.55 µm with the Ebook.

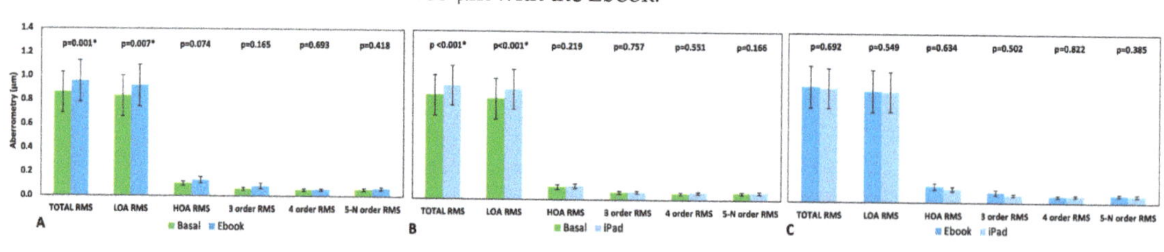

Figure 10. Comparative graphical representation of root mean square (RMS) (µm) before and after each reading (**A**,**B**) and by device after reading (**C**), considering OD and OS together. Low-order aberrations (LOAs), high-order aberrations (HOAs), 3rd-order aberrations RMS, 4th-order aberrations RMS, and 5th-order onwards (5-N) aberrations RMS. The *p* value is displayed above the graph bars, indicating * statistically significant differences; standard error is also displayed as error bars.

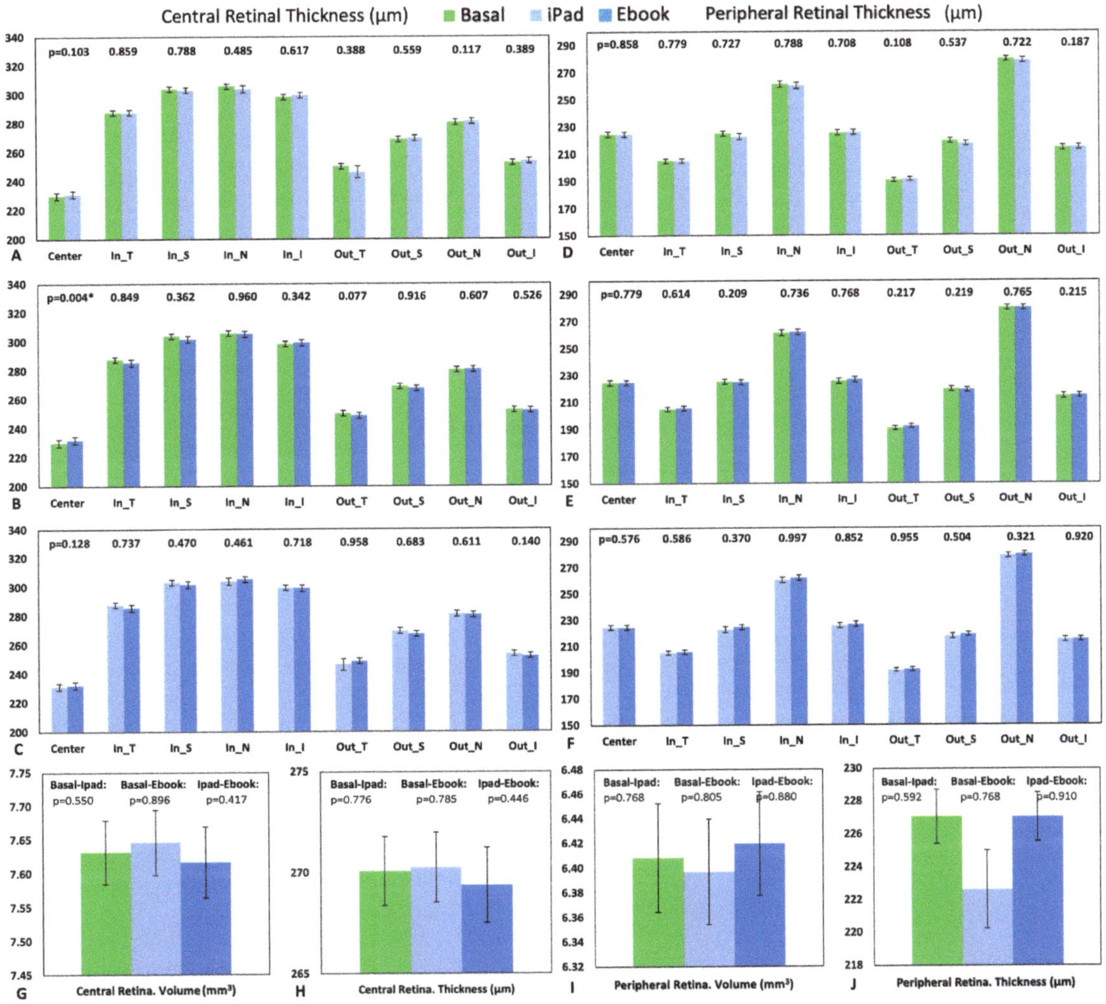

Figure 11. Graphical representation of the central retinal thickness measured by OCT: comparison of the baseline results with each device (**A**,**B**) and both devices with each other (**C**). Graphical representation of the peripheral retinal thickness measured by OCT: comparison of the baseline results with each of the devices (**D**,**E**) and between them (**F**). Graphical representation of the total central volume of the retina (**G**) and average total thickness (**H**) before the readings (baseline), after reading with the iPad, and after reading with the Ebook. Graphical representation of the total peripheral volume of the retina (**I**) and the average total thickness (**J**) before the readings (baseline), after reading with the iPad, and after reading with the Ebook. The p value is shown above the graphical bars, indicating * statistically significant differences; standard error is also displayed as error bars.

Peripheral Retina. No statistically significant changes were found in peripheral retinal thickness when comparing baseline measurements with post-reading measurements or between the two types of readings ($p > 0.05$) (Figure 11D–F). In Figure 11I, the highest total volume in the peripheral retina was observed after reading with the Ebook, with a value of 6.42 ± 0.33 mm^3, while the minimum volume was 6.40 ± 0.33 mm^3 for the iPad, with no statistically significant differences ($p > 0.05$). In the comparison of peripheral average retinal thickness between baseline measurements and post-reading measurements (Figure 11J), no statistically significant differences were found ($p > 0.05$), with the highest average thickness measured at 227.10 ± 13.10 µm in baseline measurements and the minimum average thickness at 222.62 ± 28.63 µm with the iPad, while the peripheral average retinal thickness after reading with the Ebook was 227.04 ± 11.69 µm.

4. Discussion

In this study, ocular motility was analyzed using an eye tracker through five one-minute recordings during a total reading time of 21 min using an Ebook and an iPad. Additionally, aberrometric changes were also studied to analyze the visual quality and modifications in retinal thickness after reading with each device, always under constant and comfortable ambient lighting conditions in the testing room.

Regarding fixations, as mentioned in the previous section, statistically significant differences were found between the two devices only in the average number of fixations during the first minute of reading (Figure 6A), with 187.90 ± 23.59 fixations for the iPad and 172.55 ± 26.57 fixations for the Ebook ($p = 0.002$). Overall, both the number and duration of fixations were higher during reading with the iPad. In the case of saccades, significant differences were found in their length (Figure 8A) (monocular analysis separately for each eye) in minutes 1 and 20 in the OS ($p = 0.043$ and $p = 0.020$, respectively) and in the OD at minutes 10 and 15 ($p = 0.019$ and $p = 0.024$, respectively), which could be explained by the subjective discomfort referenced by the subjects involved in our study. In contrast with our results, it has been described that parameters such as different screen refresh rates or distinct resolution do not affect saccadic eye movements or reading speed and accuracy [31,32]; this matches the results found for all other parameters such as number, duration, and velocity as there were no significant changes between the reading devices (Figures 7 and 8B).

Lighting conditions are crucial in these types of experiments. It has been observed that when both the luminance levels of the screens and the ambient lighting conditions are minimal, the total number of saccades and their duration, together with the number of blinks, are higher, which implies greater visual discomfort [26]. Poor ambient lighting conditions, characterized by uneven brightness between the screen and its background or reflections from the digital device can lead to discomfort and disabling glare, resulting in reduced contrast and a subpar image quality [33]. This diminished visual quality of electronic screens has been linked to a decrease in blink rates [34,35].

Regarding the blink rate (Figure 9), an increase in the number of blinks was observed throughout the reading with the Ebook, except at minute 20. The only significant value was at minute 1 ($p = 0.025$), indicating more blinks during reading with the Ebook, which is consistent with reading from printed paper, supporting the idea that backlit screens tend to decrease the blink rate (iPad) [36,37] and blinking abnormalities associated with changes in the ocular surface [2].

Expected values for the pupil diameter were obtained with both devices under levels of approximately 250 lux at the corneal plane. As intended, the pupil size remained constant in the monocular analysis. When comparing the two reading instruments, significant differences were found at minutes 1 and 5 ($p < 0.05$), perhaps due to the transient adaptation required to perform the experiment.

The relationship between blink frequency and tear film instability has been studied in previous research [36,37], which found that a lower blink frequency was associated with increased ocular dryness and tear film instability. Benedetto et al. [38] reported that an increased [13] blink frequency resulted in reduced tear evaporation, leading to improved tear film stability. According to Li et al. [39], not only did the subjective tear film stability increase but a higher number of saccades per second was also recorded. Although this was not observed in our study, a higher number of blinks was recorded during reading with the Ebook (Figure 9). Conversely, as observed in Figures 6A and 7A, fixations and saccades were more frequent with the iPad, and the same trend was observed in fixations for their duration, as well as the length and velocity of the saccades. It was previously described that individuals experienced poor image quality, reduced contrast or font size, potential glare, or cognitive strain in relation to computer tasks and various testing conditions. Even when using handheld electronic devices at closer distances and below eye level, individuals reported lower blink rates, which may be attributed to the angle of gaze, although the exact cause remains unknown [40]. Talens-Estarelles et al. [41] found that the blink rate remained consistent across four different types of displays: computer, tablet, e-reader, and smartphone, suggesting that the blink rate may be influenced more by cognitive demands rather than the specific display method, as mentioned by other authors [42–44].

The analysis of the results obtained with the aberrometer indicates that, in both devices under the described illumination compared to baseline measurements, there is an increase in RMS_{TOTAL}, specifically the RMS_{LOA}, with statistically significant values in the comparison of baseline measurements with each device separately ($p < 0.05$). In both the iPad and the Ebook, the aberrations found were higher than those obtained in the baseline measurements. These results confirm a transient increase in RMS_{LOA}, specifically defocus, after maintaining reading in electronic devices in young people with normal accommodation capacity.

The RMS_{TOTAL} after reading with the Ebook (Figure 10A) was 0.958 ± 1.374 µm and the RMS_{LOA} was 0.924 ± 1.376 µm, while, before the readings, they were 0.864 ± 1.340 µm and 0.836 ± 1.345 µm, respectively, with statistically significant differences ($p = 0.001$ and $p = 0.007$).

For the iPad (Figure 10B), after reading, the RMS_{TOTAL} was 0.945 ± 1.284 µm and the RMS_{LOA} was 0.918 ± 1.291 µm, while, before the readings, they were 0.864 ± 1.340 µm and 0.836 ± 1.345 µm, respectively, also with statistically significant differences ($p < 0.001$ in both cases).

However, no statistically significant differences were found for RMS_{HOA} in any of the comparisons. There were also no significant differences when comparing the values obtained between the iPad and the Ebook (Figure 10C).

There are many studies evaluating the anterior pole and digital devices [2], but the analysis of visual quality is still weak; comparisons in reading on paper versus reading on a computer screen under photopic conditions can be found [45]. In this case, after reading on the computer, individuals exhibited changes in aberrations below the 5th order. In the case of reading on paper, significant changes were observed in 3rd-order aberrations. When comparing both reading methods, no significant differences below the 5th order were found. In contrast, our study showed changes in LOA and total aberrations compared to baseline measurements. There were no statistically significant differences in aberrations between the two devices.

Regarding the retinal analysis, although the changes were not significant, it was found that when comparing central retinal thickness and volume (Figure 11G,H) and peripheral retinal thickness and volume (Figure 11I,J), the opposite phenomenon occurred. The highest thickness and volume in the central retina were observed after reading with the iPad (270.27 ± 13.54 µm and 7.65 ± 0.38 mm^3, respectively), while the maximum thickness of the peripheral retina was found in the baseline measurement (227.10 ± 13.10 µm), and the maximum peripheral volume was obtained after reading with the Ebook (6.42 ± 0.33 mm^3). When comparing peripheral retinal thickness after each reading (excluding the baseline measurement), the maximum thickness was found after reading with the Ebook. In other words, central thickening was observed after reading with the iPad, accompanied by peripheral thinning, while central thinning was observed after performing the task with the Ebook, resulting in peripheral thickening.

Although our results obtained with OCT are not statistically significant, they exhibit similar behavior in terms of central retinal thickness and volume compared to the aforementioned aberrometric results. According to another study in the literature [11], no significant differences were found in either central retinal volume or thickness; however, that study found differences in certain areas of the peripheral retina.

During reading, accommodation occurred, resulting in changes in ocular anatomy to achieve the necessary focus on the retina. This accommodation induced temporary myopia, and the data obtained in this study provide new insights into this topic. Electronic devices are considered a potential cause of myopia progression due to the sustained accommodation demand they require [46]. By maintaining maximum illumination and spatial frequency of the stimulus, axial defocus is sustained over time, stretching the retina, which is a myopiagenic factor [47], and even accommodative microfluctuations depend on the type of display used [48].

5. Conclusions

In conclusion, this study analyzed ocular motility and visual quality parameters during reading with Ebook and iPad devices. The findings demonstrated subtle variations in fixation and saccade patterns between the two devices, indicating a tendency towards increased numbers of fixations and saccades and longer durations of fixations during iPad reading, although in general not reaching statistical significance. Moreover, Ebook reading exhibited a higher blink frequency, potentially implying distinctions in tear film stability. The large errors associated with the measurement of eye motility make it difficult to say definitively whether the lack of statistical significance was due to the absence of the effect or the uncertainty of the measurements. Regarding visual quality evaluated by aberrometry, specifically, RMS_{TOTAL} and RMS_{LOA} increased significantly after reading with both devices compared to baseline measurements, suggesting a momentaneous myopization according to the sustained accommodation during reading; however, no significant differences were observed in RMS_{HOA}.

Regarding retinal thickness, in our study population, no significant changes were found in central retinal thickness between the two devices, but a slight thinning was observed after Ebook reading compared to baseline measurements and peripheral retinal thickness remained relatively stable. These results, combined with the aberrometric results, suggest that the observed changes could be attributed to anterior pole changes during accommodation remaining in the retina with a certain stability after reading for 21 min under adequate lighting.

These findings contribute to the understanding of ocular responses during reading with electronic devices and highlight the importance of considering device-specific factors when assessing visual performance and ocular health. Nevertheless, further studies are needed to explore the long-term effects of using electronic devices for reading tasks and the potential implications of such use in myopia progression.

Author Contributions: Conceptualization, M.M.-E., A.S.-C. and E.O.-H.; methodology, M.M.-E., A.S.-C. and E.O.-H.; validation, M.M.-E., A.S.-C. and E.O.-H.; formal analysis, M.M.-E., A.S.-C. and E.O.-H.; investigation, M.M.-E., A.S.-C. and E.O.-H.; resources, A.S.-C. and E.O.-H.; data curation, M.M.-E., A.S.-C. and E.O.-H.; writing—original draft preparation, M.M.-E., A.S.-C. and E.O.-H.; writing—review and editing, M.M.-E., A.S.-C. and E.O.-H.; visualization, A.S.-C. and E.O.-H.; supervision, A.S.-C. and E.O.-H. All authors have read and agreed to the published version of the manuscript.

Funding: This research was funded by the Agencia Estatal de Investigación, Ministerio de Ciencia e Innovación of the Spanish Government (grant PID2019-107058RB-I00 funded by MCIN/AEI/10.13039/501100011033), Gobierno de Aragón-Departamento de Ciencia, Universidad y Sociedad del Conocimiento-No LMP39_21, the European Union's Horizon 2020 research agreement no. 956720, and Group H07_23R: "Ayudas para la adquisición de infraestructura de investigación (Modalidad A). Contrato Programa Plan De Inversiones e Investigación Gobierno de Aragón-Universidad de Zaragoza. Fondos Feder-2020".

Institutional Review Board Statement: This study was conducted in accordance with the Declaration of Helsinki and approved by the Comité de Ética de la Investigación de la Comunidad de Aragón (CEICA) with reference PI21-074.

Informed Consent Statement: Written informed consent was obtained from the subjects involved in the study.

Data Availability Statement: Not applicable.

Conflicts of Interest: The authors declare no conflict of interest.

References

1. Moon, J.H.; Kim, K.W.; Moon, N.J. Smartphone use is a risk factor for pediatric dry eye disease according to region and age: A case control study. *BMC Ophthalmol.* **2016**, *16*, 188. [CrossRef] [PubMed]
2. Wolffsohn, J.S.; Lingham, G.; Downie, L.E.; Huntjens, B.; Inomata, T.; Jivraj, S.; Kobia-Acquah, E.; Muntz, A.; Mohamed-Noriega, K.; Plainis, S. TFOS Lifestyle: Impact of the digital environment on the ocular surface. *Ocul. Surf.* **2023**, *28*, 213–252. [PubMed]
3. Hayes, J.R.; Sheedy, J.E.; Stelmack, J.A.; Heaney, C.A. Computer use, symptoms, and quality of life. *Optom. Vis. Sci.* **2007**, *84*, E738–E755. [CrossRef]
4. Rossignol, A.M.; Morse, E.P.; Summers, V.M.; Pagnotto, L.D. Video display terminal use and reported health symptoms among Massachusetts clerical workers. *J. Occup. Med.* **1987**, *29*, 112–118.
5. López-Camones, J.J.; Rojas-Meza, L.J.; Osada, J. Frecuencia de factores ocupacionales asociados a astenopía en trabajadores usuarios de pantallas de visualización de datos de empresas del rubro construcción en Huaraz, 2019. *Rev. Asoc. Española Espec. Med. Trab.* **2020**, *29*, 56–66.
6. Ghosh, A.; Collins, M.J.; Read, S.A.; Davis, B.A.; Iskander, D.R. The influence of downward gaze and accommodation on ocular aberrations over time. *J. Vis.* **2011**, *11*, 17. [CrossRef]
7. López-Gil, N.; Fernández-Sánchez, V.; Legras, R.; Montés-Micó, R.; Lara, F.; Nguyen-Khoa, J.L. Accommodation-related changes in monochromatic aberrations of the human eye as a function of age. *Investig. Ophthalmol. Vis. Sci.* **2008**, *49*, 1736–1743. [CrossRef]
8. Wick, B.; Morse, S. Accommodative accuracy to video display monitors.: Poster# 28. *Optom. Vis. Sci.* **2002**, *79*, 218.
9. Moulakaki, A.I.; Recchioni, A.; Del Águila-Carrasco, A.J.; Esteve-Taboada, J.J.; Montés-Micó, R. Assessing the accommodation response after near visual tasks using different handheld electronic devices. *Arq. Bras. Oftalmol.* **2017**, *80*, 9–13. [CrossRef] [PubMed]
10. Wang, Y.; Shao, Y.; Yuan, Y. Simultaneously measuring ocular aberration and anterior segment biometry during accommodation. *J. Innov. Opt. Health Sci.* **2015**, *8*, 1550005. [CrossRef]
11. Orduna-Hospital, E.; Ávila, F.J.; Fernández-Espinosa, G.; Sanchez-Cano, A. Lighting-Induced Changes in Central and Peripheral Retinal Thickness and Shape after Short-Term Reading Tasks in Electronic Devices. *Photonics* **2022**, *9*, 990. [CrossRef]
12. Jaiswal, S.; Asper, L.; Long, J.; Lee, A.; Harrison, K.; Golebiowski, B. Ocular and visual discomfort associated with smartphones, tablets and computers: What we do and do not know. *Clin. Exp. Optom.* **2019**, *102*, 463–477. [CrossRef] [PubMed]
13. Benedetto, S.; Drai-Zerbib, V.; Pedrotti, M.; Tissier, G.; Baccino, T. E-readers and visual fatigue. *PLoS ONE* **2013**, *8*, e83676. [CrossRef]
14. National Reading Panel (U.S.). *Report of the National Reading Panel: Teaching Children to Read: An Evidence-Based Assessment of the Scientific Research Literature on Reading and Its Implications for Reading Instruction: Reports of the Subgroups*; National Institute of Child Health and Human Development, National Institutes of Health, University of Michigan: Ann Arbor, MI, USA, 2000.
15. Yu, H.; Akita, T.; Koga, T.; Sano, N. Effect of character contrast ratio of tablet PC and ambient device luminance ratio on readability in low ambient illuminance. *Displays* **2018**, *52*, 46–54. [CrossRef]
16. Kojima, T.; Sano, S.; Ishio, N.; Koizuka, T.; Miyao, M. Verification of the Minimum Illuminance for Comfortable Reading of an E-paper. In Proceedings of the Universal Access in Human-Computer Interaction. Applications and Services for Quality of Life: 7th International Conference, UAHCI 2013, Las Vegas, NV, USA, 21–26 July 2013; Part III 7. Springer: Berlin/Heidelberg, Germany, 2013; pp. 348–355.
17. Shen, I.-H.; Shieh, K.-K.; Chao, C.-Y.; Lee, D.-S. Lighting, font style, and polarity on visual performance and visual fatigue with electronic paper displays. *Displays* **2009**, *30*, 53–58. [CrossRef]
18. Yang, Q.; Bucci, M.P.; Kapoula, Z. The latency of saccades, vergence, and combined eye movements in children and in adults. *Investig. Ophthalmol. Vis. Sci.* **2002**, *43*, 2939–2949.
19. Epelboim, J.; Suppes, P. A model of eye movements and visual working memory during problem solving in geometry. *Vision Res.* **2001**, *41*, 1561–1574. [CrossRef]
20. Alvarez Alberdi, C.M. Técnicas de lectura eficaz. *RIFOP Rev. Interuniv. De Form. Del Profr. Contin. De La Antig. Rev. De Esc. Norm.* **1993**, *18*, 83–91.
21. Tinker, M.A. Time relations for eye-movement measures in reading. *J. Educ. Psychol.* **1947**, *38*, 1. [CrossRef]
22. Salceda, J.C.R.; Montesinos, M.M.T.; Alonso, G.A. Velocidad lectora en alumnado hispanohablante: Un metaanálisis. *Rev. Psicodidáctica* **2020**, *25*, 158–165. [CrossRef]
23. Wang, L. Test and evaluation of advertising effect based on EEG and eye tracker. *Transl. Neurosci.* **2019**, *10*, 14–18. [CrossRef] [PubMed]
24. Mele, M.L.; Federici, S. Gaze and eye-tracking solutions for psychological research. *Cogn. Process.* **2012**, *13*, 261–265. [CrossRef] [PubMed]

25. Zhang, X.; Liu, X.; Yuan, S.-M.; Lin, S.-F. Eye tracking based control system for natural human-computer interaction. *Comput. Intell. Neurosci.* **2017**, *2017*, 5739301. [CrossRef] [PubMed]
26. Orduna-Hospital, E.; Safarian Baloujeh, E.; Navarro, R.; Sanchez-Cano, A. Optical and motor changes associated with lighting and near vision tasks in electronic devices. *J. Eye Mov. Res.* **2023**, *16*, 1–15. [CrossRef]
27. Orduna-Hospital, E.; Navarro-Marqués, A.; López-de-la-Fuente, C.; Sanchez-Cano, A. Eye-Tracker Study of the Developmental Eye Movement Test in Young People without Binocular Dysfunctions. *Life* **2023**, *13*, 733. [CrossRef]
28. Orduna-Hospital, E.; Maurain-Orera, L.; Lopez-de-la-Fuente, C.; Sanchez-Cano, A. Hess Lancaster Screen Test with Eye Tracker: An Objective Method for the Measurement of Binocular Gaze Direction. *Life* **2023**, *13*, 668. [CrossRef]
29. Rovira, C. Theoretical foundation and literature review of the study of concept maps using eye tracking methodology. *Prof. De La Inf.* **2016**, *25*, 59–73. [CrossRef]
30. Global Leader in Eye Tracking for over 20 Years—Tobii. Available online: https://www.tobii.com/ (accessed on 26 June 2023).
31. Han, C.; Xu, G.; Zheng, X.; Tian, P.; Zhang, K.; Yan, W.; Jia, Y.; Chen, X. Assessing the Effect of the Refresh Rate of a Device on Various Motion Stimulation Frequencies Based on Steady-State Motion Visual Evoked Potentials. *Front. Neurosci.* **2021**, *15*, 757679. [CrossRef]
32. Nikolova, M.; Jainta, S.; Blythe, H.I.; Liversedge, S.P. Binocular advantages for parafoveal processing in reading. *Vision Res.* **2018**, *145*, 56–63. [CrossRef]
33. Liu, P.; Zafar, F.; Badano, A. The effect of ambient illumination on handheld display image quality. *J. Digit. Imaging* **2014**, *27*, 12–18. [CrossRef]
34. Chu, C.; Rosenfield, M.; Portello, J.K.; Benzoni, J.A.; Collier, J.D. A comparison of symptoms after viewing text on a computer screen and hardcopy. *Ophthalmic Physiol. Opt.* **2011**, *31*, 29–32. [CrossRef] [PubMed]
35. Pinheiro, H.M.; da Costa, R.M. Pupillary light reflex as a diagnostic aid from computational viewpoint: A systematic literature review. *J. Biomed. Inform.* **2021**, *117*, 103757. [CrossRef]
36. Divjak, M.; Bischof, H. Eye Blink Based Fatigue Detection for Prevention of Computer Vision Syndrome. In Proceedings of the IAPR International Workshop on Machine Vision Applications, Tokyo, Japan, 20–22 May 2009; pp. 350–353.
37. Blehm, C.; Vishnu, S.; Khattak, A.; Mitra, S.; Yee, R.W. Computer vision syndrome: A review. *Surv. Ophthalmol.* **2005**, *50*, 253–262. [CrossRef]
38. Benedetto, S.; Carbone, A.; Drai-Zerbib, V.; Pedrotti, M.; Baccino, T. Effects of luminance and illuminance on visual fatigue and arousal during digital reading. *Comput. Hum. Behav.* **2014**, *41*, 112–119. [CrossRef]
39. Li, J.; Song, J.; Huang, Y.; Wang, Y.; Zhang, J. Effects of different interaction modes on fatigue and reading effectiveness with mobile phones. *Int. J. Ind. Ergon.* **2021**, *85*, 103189. [CrossRef]
40. Talens-Estarelles, C.; García-Marqués, J.V.; Cervino, A.; García-Lázaro, S. Use of digital displays and ocular surface alterations: A review. *Ocul. Surf.* **2021**, *19*, 252–265. [CrossRef] [PubMed]
41. Talens-Estarelles, C.; Esteve-Taboada, J.J.; Sanchis-Jurado, V.; Pons, Á.M.; García-Lázaro, S. Blinking kinematics characterization during digital displays use. *Graefe's Arch. Clin. Exp. Ophthalmol.* **2022**, *260*, 1183–1193. [CrossRef]
42. Hou, J.; Lee, J.F.; Doherty, S. A study of the effects of mobile media on L2 text processing: Beyond offline comprehension accuracy measures. *Comput. Educ.* **2022**, *182*, 104466. [CrossRef]
43. Chen, G.; Cheng, W.; Chang, T.-W.; Zheng, X.; Huang, R. A comparison of reading comprehension across paper, computer screens, and tablets: Does tablet familiarity matter? *J. Comput. Educ.* **2014**, *1*, 213–225. [CrossRef]
44. Schwabe, A.; Lind, F.; Kosch, L.; Boomgaarden, H.G. No Negative Effects of Reading on Screen on Comprehension of Narrative Texts Compared to Print: A Meta-analysis. *Media Psychol.* **2022**, *25*, 779–796. [CrossRef]
45. Gomes, J.R.M.; de Braga Franco, S.M. Near Vision Tasks and Optical Quality of the Eye. *J. Ophthalmic Vis. Res.* **2021**, *16*, 620. [PubMed]
46. Wong, C.W.; Tsai, A.; Jonas, J.B.; Ohno-Matsui, K.; Chen, J.; Ang, M.; Ting, D.S.W. Digital Screen Time During the COVID-19 Pandemic: Risk for a Further Myopia Boom? *Am. J. Ophthalmol.* **2021**, *223*, 333–337. [CrossRef] [PubMed]
47. Landis, E.G.; Park, H.N.; Chrenek, M.; He, L.; Sidhu, C.; Chakraborty, R.; Strickland, R.; Iuvone, P.M.; Pardue, M.T. Ambient light regulates retinal dopamine signaling and myopia susceptibility. *Investig. Ophthalmol. Vis. Sci.* **2021**, *62*, 28. [CrossRef] [PubMed]
48. Hynes, N.J.; Cufflin, M.P.; Hampson, K.M.; Mallen, E.A.H. The effect of image resolution of display types on accommodative microfluctuations. *Ophthalmic Physiol. Opt.* **2022**, *42*, 514–525. [CrossRef] [PubMed]

life

MDPI

Article

Assessment of Visual Quality Improvement as a Result of Spectacle Personalization

Fruzsina Benyó [1], Lilla István [1], Huba Kiss [1], Andrea Gyenes [1], Gábor Erdei [2], Éva Juhász [1], Natalia Vlasak [3], Claudia Unger [3], Tamás Andorfi [4], Kata Réz [4], Illés Kovács [1,4,*] and Zoltán Zsolt Nagy [1]

1 Department of Ophthalmology, Semmelweis University, 1085 Budapest, Hungary; benyo.fruzsina@med.semmelweis-univ.hu (F.B.); istvan.lilla@med.semmelweis-univ.hu (L.I.); kisshuba@googlemail.com (H.K.); gyenes.andrea@med.semmelweis-univ.hu (A.G.); juhasz.eva@med.semmelweis-univ.hu (É.J.); nagy.zoltan.zsolt@med.semmelweis-univ.hu (Z.Z.N.)

2 Department of Atomic Physics, Institute of Physics, Budapest University of Technology and Economics, 1111 Budapest, Hungary; erdei.gabor@ttk.bme.hu

3 Hoya Vision Care, 1043NX Amsterdam, The Netherlands; natalia.vlasak@hoya.com (N.V.); claudia.unger@hoya.com (C.U.)

4 Department of Clinical Ophthalmology, Faculty of Health Sciences, Semmelweis University, 1088 Budapest, Hungary; andorfitamas@gmail.com (T.A.); rez.kata@semmelweis.hu (K.R.)

* Correspondence: kovacs.illes@med.semmelweis-univ.hu

Abstract: Personalized spectacles customized according to an individual's facial anatomy were developed to provide enhanced visual performance and overall comfort when compared to standard spectacles. In this comparative crossover trial, each subject was randomly assigned to wear either personalized spectacles or standard spectacles for two weeks and then tried the second pair for another two weeks. Visual acuity and reading speed were measured, and visual quality and comfort were assessed using specific questionnaires. The correlation of the wearing parameters with the subjects' satisfaction was calculated. According to our results, the subjects wearing personalized glasses reported significantly less experience of swaying and significantly higher overall satisfaction compared to those wearing the control spectacles. At the end of the study, 62% of subjects preferred the personalized spectacles, and visual quality was the primary reason for their spectacle preference followed by wearing comfort. The difference from the ideal cornea–vertex distance was significantly lower when wearing the personalized spectacles compared to the control frames. In addition, the absolute value of the difference from the ideal cornea–vertex distance was significantly correlated with patient satisfaction. These results suggest that personalized spectacles, customized according to an individual's facial anatomy for the ideal wearing parameters, result in both visual and comfort advantages for wearers.

Keywords: personalized spectacle; visual quality; progressive addition lens

check for updates

Citation: Benyó, F.; István, L.; Kiss, H.; Gyenes, A.; Erdei, G.; Juhász, É.; Vlasak, N.; Unger, C.; Andorfi, T.; Réz, K.; et al. Assessment of Visual Quality Improvement as a Result of Spectacle Personalization. *Life* **2023**, *13*, 1707. https://doi.org/10.3390/life13081707

Academic Editors: José-María Sánchez-González, Alejandro Cerviño and Carlos Rocha-de-Lossada

Received: 6 June 2023
Revised: 4 August 2023
Accepted: 6 August 2023
Published: 8 August 2023

1. Introduction

Proper corrective glasses play a vital role in our modern lifestyle. With the increasing reliance on digital devices and the prevalence of near work activities, such as reading, writing, and using computers, our eyes are constantly under strain. Corrective glasses, tailored to our specific visual needs, help alleviate the discomfort and potential eye strain caused by refractive errors and presbyopia during quick adaptation to different distances when using computers, tablets, and smartphones [1]. Corrective lenses include distance glasses, reading glasses, bifocals, and progressive addition lenses, [2] the latter of which provide the desired additional power by continuously increasing the dioptric power between the distance and near zones of the lens [3–7]. It is known that some presbyopes wearing progressive addition lenses experience moderate-to-severe visual [8] symptoms, such as blurred vision, headaches, peripheral visual field distortion, imbalance, and even nausea [9], which eventually lead to the discontinuation of progressive addition lenses. The reason why some presbyopes find it so difficult to adapt to progressive addition lenses, while others find them easy to wear, is not fully understood [10,11]. The spectacle lens industry has been working for many decades to improve the design of spectacle lenses in order to meet this increasing demand for comfortable vision [12,13]. An early type of progressive addition lens was produced based on standardized wearing parameters and was a symmetrical design that showed typical side effects, such as headaches, a narrowed field of vision, a swimming/rolling sensation, discomfort, or feeling unsafe on stairs [14]. Initially, a long period of adaptation and high motivation was required from the wearers [15,16]. Even after a long period of adaptation, some wearers rejected the spectacles because they did not adapt. For these subjects, the inconvenience and lack of visual comfort and performance outweighed the benefits of perfect vision at all distances [17].

As shown previously, high visual performance is strongly linked to how a lens is positioned in front of the eyes [18], and thus, theoretically, individualized spectacles could provide better acceptance by both monofocal and multifocal spectacle wearers because of the improved visual comfort and performance offered [19]. This performance enhancement is currently limited by standardized spectacle frames as they only allow for the adaptation to anatomical and morphological facial requirements in a certain range. In the past few years, a new generation of mono- and multifocal lens design has been developed considering, for the first time, multiple individual spectacle frame parameters, such as wearers' pantoscopic angle (WPA), frame face form angle (FFA), frame–cornea distance (FCD), and frame-related centration data. Initially manually determined, the so-called video-centration measurement systems soon made it possible to measure the frame-related parameters directly on the worn frame more precisely. At the same time, to achieve the optimal lens position for the individual, the frame should also be individually tailored.

The purpose of this study is to assess the visual performance and wearing comfort of personalized spectacles produced by taking into account the individual facial parameters of a wearer compared to data obtained from wearing spectacles using a standard pre-produced frame.

2. Materials and Methods

This comparative study was conducted at the Department of Ophthalmology, Semmelweis University, Budapest, Hungary. The study followed the tenets of the Declaration of Helsinki and was approved by the Ethics Committee of Semmelweis University (29819/AOSZE/2018) before commencement. After the protocol had been fully explained, all subjects provided written informed consent to participate in the study spontaneously and for free and gave permission to collect and examine their personal and optometric data.

2.1. Eligibility Criteria

Altogether, 60 subjects were enrolled in this study. The subjects were experienced wearers wearing a single vision/progressive addition lens design during the previous six months with the best-corrected monocular visual acuity for distance and near $\geq$ 1.0 decimal. The participants required a new eyeglass prescription and had normal binocular vision and ametropia between -6.00–$+6.00$ spherical diopter and a cylindrical diopter $\leq$ 2.50D. The difference in the power (spherical equivalent) between the eyes was $\leq$2.00 D. A detailed description of the inclusion and exclusion criteria can be found in Table A1.

In total, 30 subjects wore a single vision lens design (SV group) and 30 subjects were prescribed a progressive addition lens design (PAL group).

2.2. Study Design

Each subject was randomly assigned to wear either the personalized spectacles or the standard spectacles for a period of two weeks. The subjects then tried the second pair (crossover) for another period of two weeks. The spectacles were dispensed in an unmarked box. During the study, the investigators who performed the measurements and the subjects were blind to which spectacles were worn by which subjects, so it was impossible to differentiate between the personalized and standard spectacles.

2.3. Study Devices

Two types of spectacles were worn during the study: the personalized spectacles and the control spectacles, which were visually indistinguishable; both had identical material, color, and shape. All personalized spectacles were produced using HOYA's patented YUNIKU technology.

2.4. Personalized (3D Tailored) Spectacles

YUNIKU spectacles consist of a personalized frame, produced on demand for each subject, and standard lenses in the ideal position. Before producing the YUNIKU spectacles, the subjects' faces were scanned by a YUNIKU scanner, and 3D images of their faces were rendered by YUNIKU software. Prescription and satisfaction levels with previous spectacles, functional requirements according to the spectacles' usage, and the facial reference points of each subject were collected by the YUNIKU software. The algorithm used these data to calculate the ideal position of the lens. Taking the lens' position into account, a virtual frame model is adjusted to the calculated lens parameters and to the morphology of the face. The frame is produced on demand for each subject using laser sintering technology, which is an additive manufacturing process that uses a high-powered laser to fuse powdered materials together to create three-dimensional objects. This technology offers several advantages over traditional manufacturing methods of spectacle frames as it allows for the creation of individual geometries that may be difficult or impossible to achieve with traditional manufacturing techniques. The lens position is designed corresponding to the calculated individual lens-wearing parameters (Figure 1). Both the YUNIKU single vision lens position and the YUNIKU progressive addition lens position were determined in a 3D design space.

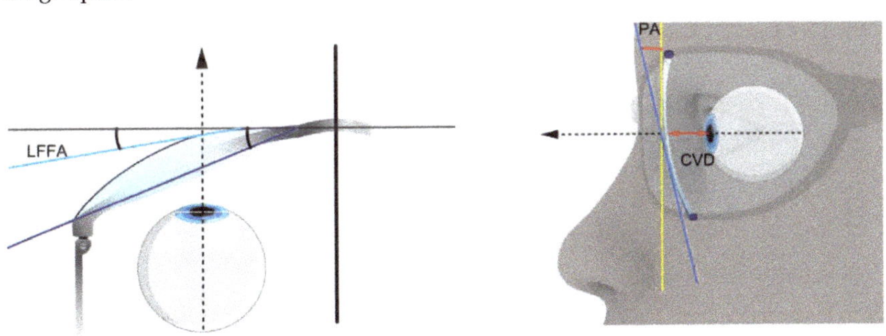

Figure 1. Individual lens-wearing parameters: CVD: cornea–vertex distance; PA: pantoscopic angle; and LFFA: lens face form angle.

2.5. Control Standard Spectacles

The frame of the control spectacles is pre-produced in one standard fixed size. The frame size according to the boxing system and bridge is adapted to facial needs individually in steps of 1 mm. The frame has the same look, material, color, and shape as the individual YUNIKU frame. The same aspherical single vision lens design and progressive addition lens design were used for both the control spectacles and the YUNIKU frames. The frame-related parameters were measured by using a video-centration system called visuReal portable (Figure 2).

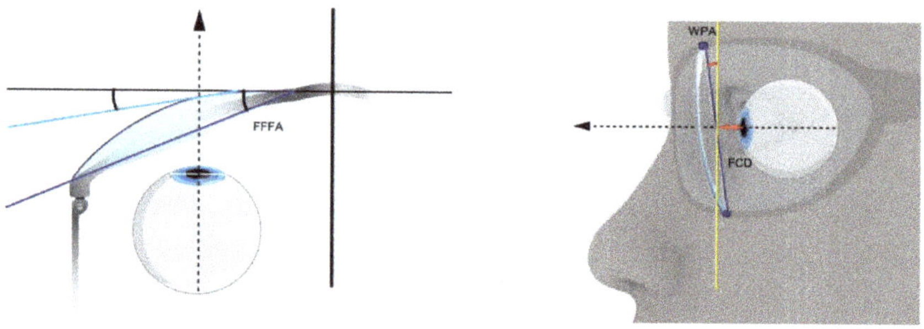

Figure 2. Frame-related wearing parameters: FCD: frame–cornea distance; WPA: wearers' pantoscopic angle; and FFFA: frame face form angle.

Subjects with inadequate fitting of the control frame according to the standard care of prescribing spectacles by a trained optometrist were excluded from the study. All YUNIKU and control lenses were verified by the binocular eye model [20] to ensure seamless matching with the visual system, resulting in supreme depth perception. During the production process, each progressive addition lens is corrected by binocular harmonization technology (BHT) leading to better intermediate and near vision. BHT recalculates and adjusts the progressive power distribution according to the actual used positions on each lens so that both eyes will experience the same accommodation support. The main differences between the construction of personalized spectacles and control standard spectacles are summarized in Table 1.

Table 1. Comparison of the construction characteristics of the personalized and control spectacles.

Criteria	Personalized Spectacles	Control Spectacles
Manufacturing	Personalized frame, produced on demand for each subject	Pre-produced frame based on standard parameters in one fixed size
Measurement and adjustment to the subject's face topography	Face scan and 3D rendering of the subject's face Calculation of the ideal lens parameter/position based on prescription, functional requirements, and facial data	Manual adjustment of the frame to the subject's face The video-centration system visuReal portable is used to measure all frame-related wearing parameters
Definition of frame size and position	The parametric frame model is adjusted to the calculated lens position and face topography	Selection of a frame that fits as best as possible
Lens design	Ideal lens-related parameters are considered	Individual frame-wearing parameters are converted into lens-related parameters

2.6. Outcomes

The primary outcome of the study was to identify a preference for either the personalized spectacles or the control standard spectacles at the end of the study. During the study, visual quality and spectacle comfort were evaluated by specific questionnaires assessing patients' first impression (Table A2), subject satisfaction (Table A3), and final comparison (Table A4) with responses to questions scored 0–5 or 1–10 as indicated. Secondary outcomes presented in this article include the evaluation of visual performance by the 25-item Visual Function Questionnaire (VFQ-25) and by measuring the reading speed by using a standard Radner test. The Radner Reading Charts consist of standardized "sentence optotypes" that logarithmically progress in print size and were designed for clinical and research use. This test provides a number of different reading parameters from a single examination in patients with normal-to-low vision, and its reliability and validity have already been demonstrated [21,22].

2.7. Statistical Analysis

A statistical analysis was performed by using SPSS software (version 23.0, IBM, Armonk, NY, USA). The sample size was determined a priori by statistical power calculation (power 0.80; $p = 0.05$), and the minimum number of patients to enroll in this study was calculated to be 58. The Shapiro–Wilk W-test was used to test the normality of the data. Due to the non-normal distribution of the data, the Wilcoxon signed rank test for the dependent samples was used to analyze the differences between the scores of different subsets of questionnaires obtained from the two different spectacles. The Chi-square test was used to test relationships between categorical variables. The binomial test with confidence intervals was used to analyze the preferences of frame selection. Principal component analysis using Varimax rotation was conducted to examine the factor structure of the final comparison questionnaire. Principal component analysis allows a large number of variables to be condensed into relatively few new variables, or principal components, that give the most information about the data. These principal components were then analyzed to determine which ones correlated with the spectacle preference. In all statistical analyses, a *p*-value of less than 0.05 was considered to be statistically significant.

3. Results

None of the 60 enrolled subjects (female: 37; male: 23) was excluded from the study, and all subjects completed all visits. All patients had a best-corrected distance visual acuity and best-corrected near visual acuity of 1.0 (i.e., 0.0 logMAR) or better. The subjects' mean sphere refraction error was -1.58 ± 2.51 diopter in the SV group and -0.34 ± 2.55 diopter in the PAL group ($p = 0.03$). The cylindrical error was -0.49 ± 0.66 diopter in the SV group and -0.45 ± 0.41 diopter in the PAL group ($p = 0.86$). Patients in the PAL group required an average of 2.17 ± 0.43 addition in their progressive addition lenses. The fitting properties of the spectacles are presented in Table 2.

Table 2. Fitting properties of the personalized and control spectacles. The standard parameters of traditional prescription eyewear for progressive addition lenses (PAL) and single vision lenses (SV) are also indicated.

		Lens Designs					
		PAL			SV		
		Personalized	Control	Standard	Personalized	Control	Standard
Vertex Distance [mm]	CVD FCD	13.82 ± 2.51	16.25 ± 3.77	12.30	13.76 ± 3.11	14.91 ± 3.07	12.30
Pantoscopic Angle [°]	PA WPA	8.75 ± 2.38	8.62 ± 4.27	8.30	4.04 ± 2.78	8.83 ± 2.87	8.30
Face Form Angle [°]	LFFA FFFA	1.14 ± 0.45	1.75 ± 1.40	4.40	1.24 ± 0.24	1.31 ± 0.71	4.40

Note: CVD: cornea–vertex distance, FCD: frame–cornea distance, PA: pantoscopic angle, WPA: wearers' pantoscopic angle, LFFA: lens face form angle, and FFFA: frame face form angle. Data: mean ± standard deviation.

3.1. Evaluation of Visual Functions and Reading Speed

There was no difference in the VFQ-25 scores when wearing the personalized spectacles compared to wearing the control spectacles (93.29 ± 6.01 vs. 93.27 ± 5.87; $p = 0.82$). However, the subjects showed significantly higher reading speeds (less time to read the Radner chart) when wearing the personalized spectacles in comparison to when wearing the control spectacles (5.73 ± 1.05 s vs. 5.79 ± 1.01 s; $p = 0.04$).

3.2. First Impression Questionnaire

According to the results of the first impression questionnaire, there was no significant difference between the personalized and control spectacles in any items related to visual quality or comfort provided by the spectacles (Table A5).

3.3. Subject Satisfaction Questionnaire

Overall satisfaction with each pair was assessed using the satisfaction questionnaire, which was completed after 14 days of wearing the new spectacles. Significantly higher scores were reported in distance (4.65 ± 0.68 vs. 4.25 ± 1.08; p = 0.009) and near vision (4.48 ± 0.95 vs. 4.13 ± 1.17; p = 0.01) by subjects wearing the personalized spectacles (Table A3). In addition, the subjects reported significantly less experience of swaying (4.40 ± 0.92 vs. 4.00 ± 1.18; p = 0.02) and significantly higher overall satisfaction (4.13 ± 0.91 vs. 3.73 ± 1.15; p = 0.03) when wearing the personalized spectacles (Table A6).

3.4. Final Comparison Questionnaire

There was no statistically significant difference in the answer scores in the final comparison questionnaire, which was completed during the last visit after the subjects tried both spectacles. As a primary outcome of the study, 37 (62%) out of 60 subjects chose the personalized spectacles as the ones they would keep. This preference towards the personalized spectacles was statistically significant (p = 0.04; Table 3). However, there was no statistically significant preference for the personalized spectacles when analyzing the SV and PAL groups separately (p > 0.05).

Table 3. Scores of the final comparison questionnaires in the single vision (SV) and progressive addition lens (PAL) groups.

Question	SV Group (n = 30)			PAL Group (n = 30)			All spectacles (n = 60)		
	Personalized	Control	p	Personalized	Control	p	Personalized	Control	p
How comfortable are the spectacles in general? (1–5)	4.03 ± 0.93	3.67 ± 1.37	0.30	4.30 ± 0.92	4.03 ± 0.81	0.11	4.17 ± 0.92	3.85 ± 1.13	0.13
How easily and quickly could you adapt to them? (1–5)	3.87 ± 1.14	3.97 ± 1.40	0.71	4.07 ± 1.01	3.80 ± 1.21	0.33	3.97 ± 1.07	3.88 ± 1.30	0.86
How comfortable are the spectacles on your nose? (1–5)	3.77 ± 1.25	3.47 ± 1.43	0.44	4.17 ± 1.05	4.10 ± 0.96	0.34	3.97 ± 1.16	3.78 ± 1.25	0.37
How comfortable are the temples? (1–5)	4.30 ± 1.09	3.73 ± 1.44	0.11	4.27 ± 0.79	4.17 ± 0.79	0.35	4.28 ± 0.94	3.95 ± 1.17	0.10
How satisfied are you with your vision in general? (1–5)	4.77 ± 0.50	4.50 ± 0.97	0.19	4.37 ± 0.89	4.17 ± 1.18	0.36	4.57 ± 0.74	4.33 ± 1.08	0.15
How satisfied are you with the visual field? (1–5)	4.47 ± 0.68	4.33 ± 1.06	0.59	4.30 ± 0.99	4.13 ± 1.04	0.37	4.38 ± 0.85	4.23 ± 1.05	0.28
Did you experience a swaying feeling when wearing the glasses? (1–5)	4.40 ± 0.93	4.23 ± 1.17	0.46	4.27 ± 1.01	3.87 ± 1.20	0.38	4.33 ± 0.97	4.05 ± 1.18	0.06
Please rank the spectacles (1–5)	8.00 ± 1.76	7.27 ± 2.72	0.44	7.97 ± 1.81	7.40 ± 2.46	0.39	7.98 ± 1.80	7.33 ± 2.57	0.18
Which pair of spectacles would you like to keep?	17	13	0.29	20	10	0.40	37	23	0.04
OVERALL SCORE	37.6 ± 6.17	35.17 ± 9.79	0.52	37.73 ± 6.09	35.57 ± 7.55	0.41	37.67 ± 6.08	35.37 ± 8.67	0.17

Note: Patient satisfaction was assessed linearly on a 1–5-point scale, with 5 indicating the best quality of vision or comfort and 1 indicating the worst. p: Wilcoxon signed-rank test and binomial test with confidence intervals. Data: mean ± standard deviation.

When analyzing the overall satisfaction after wearing both of the studied spectacles, the subjects tended to be more satisfied with the personalized spectacles (Figure 3).

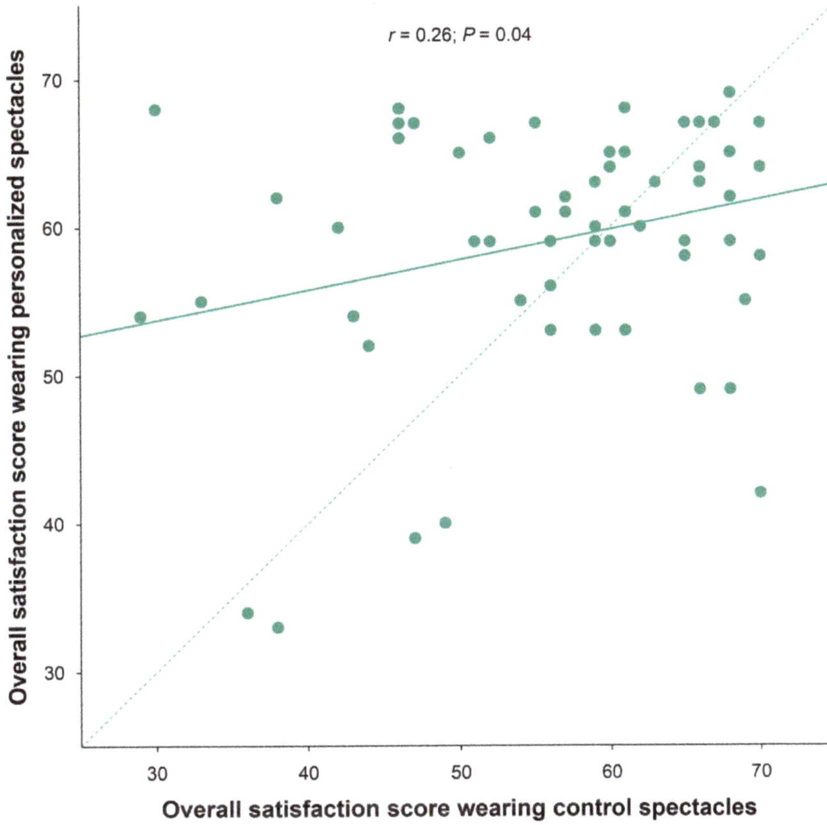

Figure 3. Correlation of overall satisfaction scores between the two study spectacles. Each circle represents a specific subject.

The principal component analysis identified two factors that accounted for 75.93% of the variance in the final comparison questionnaire scores. The first component (visual quality) explained 54.45% of the variance and was primarily responsible for the spectacle preference, while the second component (spectacle comfort) only accounted for 21.48% of the variance in scoring and had a secondary role in spectacle preference.

Finally, we found a statistically significant difference between the personalized and control spectacles in terms of deviation from the ideal cornea–vertex distance values (12.5 mm), which is the design target of the spectacle lenses, in order to keep the spectacle magnification around unity. This value was set in the trial frame during refraction. In the case of the personalized frames, the mean difference was much lower than for the control frames (2.37 ± 2.05 mm vs. 4.48 ± 2.85 mm; $p = 0.001$, Figure 4), which clearly shows the effectiveness of frame customization.

In addition, the correlation analysis showed that the absolute value of the difference (in mm) from the ideal cornea–vertex distance value for the personalized spectacles was significantly correlated with patient satisfaction ($p = 0.01$). The correlation coefficient resulted in being $r = -0.31$, implying that a decreasing position error increases patient satisfaction (Figure 5).

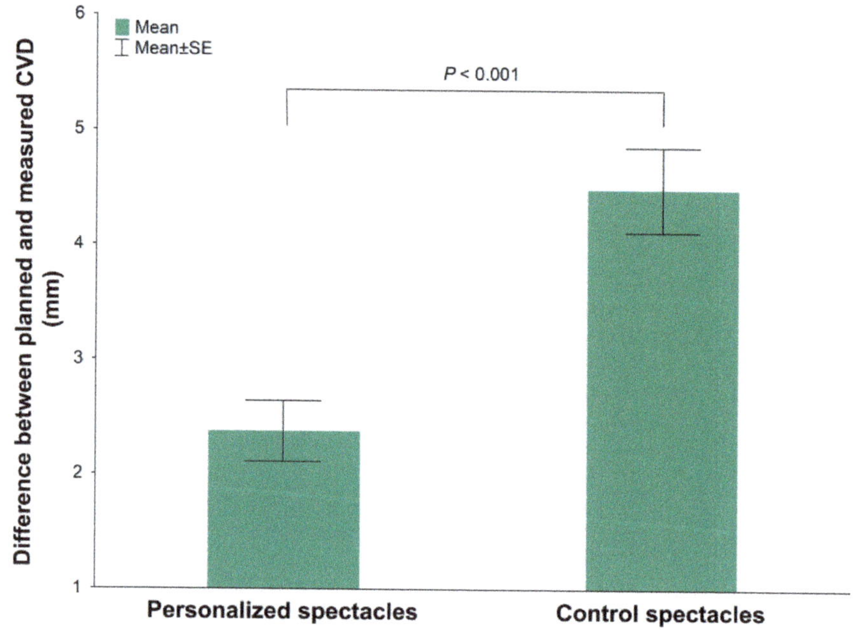

Figure 4. Absolute value of the difference in the measured CVD from the planned values when wearing the personalized and control spectacles. Data: mean ± standard error.

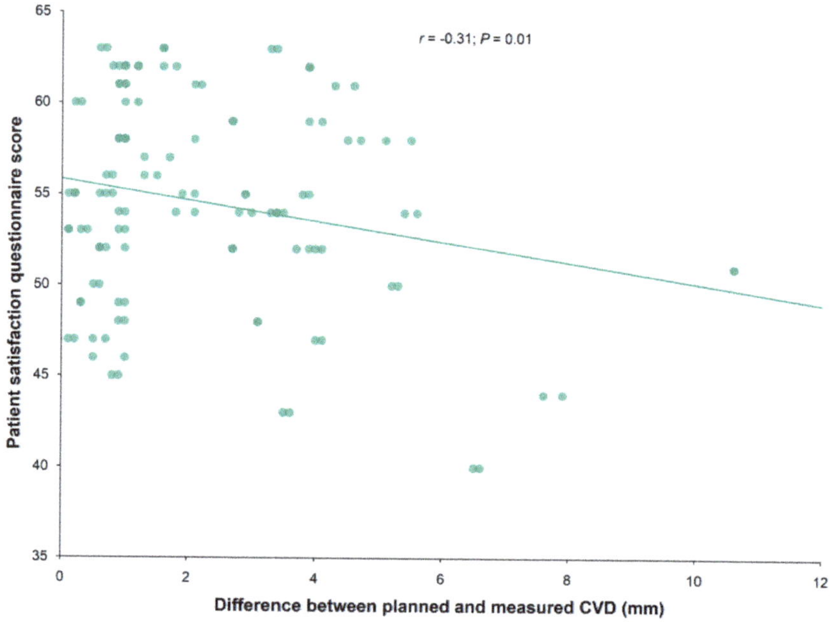

Figure 5. Correlation between deviation from the ideal CVD and patient satisfaction.

Regarding the other parameters, the lens face form angle's differences from the ideal value (0°) did not show a similar effect on patient satisfaction ($p = 0.59$), while the pantoscopic angle had no specific ideal value at all. The ideal LFFA was the design target of the spectacle lenses, defined as LFFA = 0°, since the human eye looks straight forward on average; thus, its optical axis can be considered as lying in the vertical plane of the head. PA is different for each facial structure and visual situation (reading, watching TV, driving, etc.); this is why an ideal position cannot be defined.

4. Discussion

In this clinical study, we have demonstrated that although there was no immediate benefit after putting on personalized spectacles when wearing them for a long time, the subjects were more satisfied with distance and near vision, reported less swaying, and were more satisfied in general compared to subjects wearing the control spectacles. After completing this crossover comparative study, we have shown that a significantly larger proportion of subjects preferred the personalized spectacles over the control spectacles, suggesting that improved visual performance and overall comfort while wearing the personalized spectacles is noticeable even when the best-corrected visual acuity is similar with these two spectacles. By using principal component analysis of the patient satisfaction questionnaires, we have shown that visual quality was the primary component that correlated with the preference to choose personalized spectacles. Spectacle comfort, the second principal component, was uncorrelated with visual quality and was found not to be of major importance in spectacle preference. Although the difference between the personalized and control spectacles in terms of subjective experiences was relatively small, patient satisfaction with the personalized spectacles was dependent on deviation from the ideal cornea–vertex distance, thus demonstrating the clinical relevance of proper lens positioning to deliver the best visual quality. The deviation of the actual cornea–vertex distance from the ideal cornea–vertex distance was significantly lower for the personalized frames. The level of satisfaction was proportional to this deviation, as the closer the cornea–vertex distance was to the optimal value, the higher the level of satisfaction reported by the participants. An interesting observation was that people who gave a lower score for comfort when wearing standard glasses had a greater improvement in satisfaction scores when wearing customized glasses. Although the reasons for this finding were not investigated in this study, it may be due to the more unusual face form of these people, which makes it more difficult to fit standard spectacles correctly.

The results of this study are particularly important for progressive addition lens wearers, who usually experience some discomfort symptoms related to the progressive addition lens design [23]. Since many factors may lead to discomfort [24], such as the effect of a small reading area on depth perception and contrast sensitivity [25,26] or a narrow corridor due to the large amounts of astigmatism at the edges, eye care professionals do not have a standardized technique to determine which patients will likely adapt to progressive addition lenses and which patients will have a difficult time. The design and width of the progressive addition lens channel can vary depending on the specific lens manufacturer and the wearer's prescription needs. Some progressive addition lenses may have wider channels, providing a larger area for near-vision tasks, while others may have narrower channels that prioritize distance vision. It is also known that ear position has a significant effect on pantoscopic angle, resulting in an unwanted tilt of spectacle lenses leading to visual disturbances or a long period of adaptation to progressive addition lenses [25,26]. The pantoscopic angle is defined not only by the position of the wearers' ears on their heads in relation to their eyes but also by the predefined inclination of the spectacle frame, which can be slightly adjusted. In order to compensate for a pantoscopic angle that is not ideal, the progressive lens design is converted to that angle. However, even though the technology for modern lens design is much more enhanced than 60 years ago, there are still cases of non-adaptation.

Wearing the correct progressive addition lenses is particularly important for people working on computers as screen time, including smartphone use, has increased significantly in Western countries over the past decade. In 2020, adults in the United States spent an average of 13 h a day interacting with media, including screen time on computers, smartphones, tablets and TVs. This is a significant increase from 10 years ago, when average daily screen time was around 7 h. Advances in technology have also increased the resolution of such devices: smartphones with more than 300 dpi and UHD/4K desktop monitors have been with us for more than a decade, both of which challenge the eye's resolution capacity. Improved visual quality is therefore increasingly important, as it can have a direct impact on the effectiveness of computer work. When people have clear and sharp vision, they can read text, view images and interpret visual data with greater accuracy and efficiency. This reduces the time spent deciphering unclear or blurred information, leading to improved productivity and task completion. In addition, improved visual quality reduces eye strain and fatigue, allowing people to work comfortably for longer periods of time without experiencing visual discomfort. It is also important to note that inappropriate spectacle correction can aggravate dry eye symptoms, leading to increased discomfort and visual disturbance.

The best possible position of corrective spectacles on the subject's face is a prerequisite for quick adaptation and the best optical quality provided by the spectacles [27]. The fitting of a frame by an optometrist is a collaborative process that involves understanding the individual's facial anatomy, prescription requirements, and personal preferences. During a frame fitting, the optometrist considers various factors to determine the most suitable frame for the individual. They take into account facial dimensions, such as interpupillary distance and facial width, to ensure proper alignment of the lenses with the individual's eyes. The optometrist also considers the shape of the individual's face and features to recommend frames that complement their facial structure and personal style. Additionally, the optometrist assesses the position of the ears and the shape of the nose to ensure a comfortable and secure fit. They make necessary adjustments to the frame, such as temple length and nose pad positioning, to customize the fit for the individual. The optometrist also considers the individual's prescription needs, ensuring that the frame can accommodate the required lens thickness and curvature. By taking into account these factors, the optometrist ensures that the selected frame provides the best possible visual experience and meets the individual's unique needs. However, new technologies allow many more options for personalization as biometric data can be captured quickly and with high resolution, using a 3D scanner and personalized manufacturing production methods, such as the laser sintering process [28–30]. Instead of taking standardized spectacle frames with their limitations as given and accepting the associated constraints in visual performance and wearing comfort, new technologies can be used in such a way that allows personalized spectacles to optimize the position of the lens in front of the eye, providing a higher level of visual performance and wearing comfort.

The personalized spectacles, known as YUNIKU, are considered a completely new approach to fully personalized spectacles, as the YUNIKU software calculates the ideal position of the lens in relation to the eyes and then designs the frame based on those unique parameters. Laser sintering technology allows tailoring of the frame to optimize visual, aesthetic, and comfort performance, and lens design does not need to be reverse-adjusted based on frame-related parameters. According to the results of this clinical study, this level of spectacle personalization provides slightly better visual performance and satisfaction than the control spectacles. In this study, we have shown that reading speed was also slightly better when wearing personalized spectacles. Although the difference was minimal, faster reading might be attributed to the less compromised visual quality due to more optimal lens tilt in the personalized spectacles. The significance of this finding is that previous research on people working at a computer screen has shown that increasing reading speed can lead to improved comprehension, increased productivity and reduced eye strain. As unwanted eye strain due to lens misalignment is known to increase with increasing prescription, a more pronounced benefit in visual performance and comfort may be found by wearing customized frames in subjects with high ametropia compared to the relatively small differences found in this study. Since visual demands, especially for near vision, are becoming increasingly more important in the workplace, there is an increasing demand for improved spectacle comfort, especially for progressive addition lens wearers. However, it is still unknown as to why some wearers adapt very quickly to these lenses while others complain of headaches, swaying effects, and distorted peripheral vision as well as experiencing problems in the workplace. It has already been demonstrated that vergence facility and the rate of phoria adaptation may have potential clinical utility in differentiating which patients may adapt to progressive addition lenses and which ones will have more difficulty.

There are some limitations to this study. First, there is a possible effect of different spherical equivalent values and cornea–vertex distance values measured in the single vision and the progressive addition lens groups on the study outcomes. However, this study has shown that wearing personalized spectacles manufactured with the ideal wearing parameters improved the subjects' visual performance and overall satisfaction when analyzing the whole cohort. The lack of a statistically significant preference for personalized spectacles when analyzing the SV and PAL groups separately is probably a consequence of the fact that there is little difference between the results when wearing personalized and standard spectacles. Second, the relatively small sample size may make it difficult to determine if a particular outcome is really a true finding, and in some cases, a type I error may occur, especially in the case of multiple comparisons. We recommend further studies with larger sample sizes, which should lead to more reliable conclusions. Another limitation of this study was, that the same aspheric single-vision lens design and progressive addition lens design were used for both the control spectacles and the YUNIKU frames. While this helps to isolate the difference in this study caused by the difference in frames, if the lenses had been customized for the control frames, as is common with many lens manufacturers nowadays, there might have been less of a subjective difference. Finally, it is also possible that if subjects who reported no benefit of personalized progressive addition lenses were to wear progressive addition lenses longer, then these subjects might have eventually shown a preference for personalized progressive addition lenses. However, it is beyond the scope of this study to determine whether subjects who self-reported that they could not choose between personalized and control progressive addition lenses after 1 month would eventually prefer personalized progressive addition lenses if they were given more time to wear them. Further studies on a patient cohort with a higher level of ametropia is suggested to identify the plausible relationship between dioptric power and the probability of non-adaptation to progressive addition lenses.

5. Conclusions

This study has shown that wearing personalized spectacles manufactured according to an individual's facial anatomy with the ideal wearing parameters results in some advantages. Slightly improved visual performance and comfort is noticeable while wearing the personalized spectacles.

Author Contributions: Conceptualization, I.K., K.R., T.A. and Z.Z.N.; methodology, I.K.; formal analysis, I.K.; investigation, F.B., A.G., H.K., A.G., G.E. and É.J.; resources, Z.Z.N.; data curation, F.B., A.G., H.K., G.E. and T.A.; writing—original draft preparation, F.B, L.I., I.K., A.G., G.E., H.K, É.J. and Z.Z.N.; writing—review and editing, T.A. and K.R.; supervision, Z.Z.N.; project administration, I.K., N.V. and C.U. All authors have read and agreed to the published version of the manuscript.

Funding: The APC was funded by Semmelweis University, Budapest, Hungary.

Institutional Review Board Statement: The study was conducted in accordance with the Declaration of Helsinki and approved by the Institutional Ethics Committee of SEMMELWEIS UNIVERSITY (29819/AOSZE/2018).

Informed Consent Statement: Informed consent was obtained from all subjects involved in the study.

Data Availability Statement: Data are available from the authors upon reasonable request.

Conflicts of Interest: Natalia Vlasak and Claudia Unger are employees of Hoya Vision Care, the Netherlands. The other authors declare no conflicts of interest. The funders had no role in the design of the study; in the collection, analysis, or interpretation of the data; in the writing of the manuscript; or in the decision to publish the results.

Appendix A

Table A1. Inclusion and exclusion criteria for the study participants.

Inclusion Criteria	1. Age: - Single vision design (SV group): 18–44 years - Progressive addition lens design (PAL group): 45–65 years 2. The subjects should be experienced wearers wearing SV/PAL designs during the last six months. 3. Visual Acuity: distance and near monocular $\geq$ 1.0 ($\leq$0.0 logMAR) 4. Normal binocular vision 5. Ametropia: - Spherical power: −6.00–+6.00 D - Cylindrical power: $\leq$ 2.50 D - Difference in power (spherical equivalent) between TE eyes: $\leq$ 2.00 D
Exclusion Criteria	1. First prescription for progressive addition lenses 2. The prescription varies from the previous prescription by more than 0.75 D in the spherical equivalent or in the cylinder axis by more than 15° in any eye 3. Double vision or prismatic power in current glasses 4. Known ocular disease including strabismus, any pathology, and any eye surgeries that may affect visual acuity 5. The use of systemic or ocular medication that is likely to affect vision 6. Balance problems/vertigo problems 7. Concurrent participation in any other vision-related studies 8. Pregnancy 9. Inadequate fitting of control frames

Table A2. First impression questionnaire.

Question
How comfortable are the new spectacles in general? (1–10)
How comfortable are the new spectacles on your nose? (1–10)
How do you find the spectacles' bridge on your nose?
1: There is a space between the bridge and my nose
2: It presses on my nose
3: It fits perfectly on my nose
How are the temples? (1–10)
How do you find the temples according to your head?
1: They are too loose on my head
2: They press on my head
3: They fit perfectly according to my head
How is the length of the temples?
1: Too long
2: Too short
3: The temple's length is exactly what I need
How is your vision in general with the new spectacles? (1–10)
How do you find the field of view with the new spectacles in general? (1–10)
Do you notice any distortion of images or blurring?
1: No, not at all
2: Slightly
Do you experience any swaying feelings?
1: No, not at all
2: Slightly
OVERALL SCORE

Note: Patient satisfaction was assessed linearly on a 1–10-point scale, with 10 indicating the best quality of vision or comfort and 1 indicating the worst.

Table A3. Self-evaluation questionnaire.

Question
How comfortable is this pair of spectacles in general? (1–5)
How easily and quickly could you adapt to this pair of spectacles? (1–5)
Spectacles slide down from the nose (1–5)
Lenses steam up while wearing them (1–5)
How comfortable was this pair on your nose during the two weeks of wearing them? (1–5)
How comfortable were the temples during the two weeks of wearing this pair? (1–5)
How satisfied are you with your vision/visual acuity using this pair looking:
At far distances (>4–6 m: TV. . .)? (1–5)
At intermediate distances (≈60 cm–1 m: computer work) (1–5)
Doing your near work (≈30–50 cm: reading...) (1–5)
How satisfied are you with the visual field using this pair:
At far distances (>4–6 m: TV. . .)? (1–5)
At intermediate distances (≈60 cm -1 m: computer work) (1–5)
Doing your near work (≈30–50 cm: reading...) (1–5)
Did you experience any swaying feeling using this pair? (1–5)
How satisfied are you with this eyewear in general? (1–5)
OVERALL SCORE

Note: Patient satisfaction was assessed linearly on a 1–5-point scale, with 5 indicating the best quality of vision or comfort and 1 indicating the worst.

Table A4. Final comparison questionnaire.

Question
How comfortable are the spectacles in general? (1–5)
How easily and quickly could you adapt to them? (1–5)
How comfortable are the spectacles on your nose? (1–5)
How comfortable are the temples? (1–5)
How satisfied are you with your vision in general? (1–5)
How satisfied are you with the visual field? (1–5)
Did you experience a swaying feeling when wearing the glasses? (1–5)
Please rank the spectacles (1–5)
Which pair of spectacles would you like to keep?
OVERALL SCORE

Note: Patient satisfaction was assessed linearly on a 1–5-point scale, with 5 indicating the best quality of vision or comfort and 1 indicating the worst.

Table A5. Results of the first impression questionnaire in the single vision (SV) and progressive addition lenses (PAL) groups.

	SV Group (n = 30)			PAL Group (n = 30)			All spectacles (n = 60)		
Question	Personalized	Control	p	Personalized	Control	p	Personalized	Control	p
How comfortable are new spectacles in general? (1–10)	8.70 ± 1.23	7.46 ± 2.55	0.03	8.40 ± 1.59	8.43 ± 1.53	0.54	8.55 ± 1.42	7.95 ± 2.26	0.19
How comfortable are the new spectacles on your nose? (1–10)	8.23 ± 1.92	7.30 ± 2.84	0.20	8.50 ± 1.19	8.87 ± 1.04	0.14	8.37 ± 1.59	8.08 ± 2.26	0.94
How do you find the spectacles' bridge on your nose									
1: There is a space between the bridge and my nose	2	8	0.65	5	3	0.31	7	11	0.54
2: It presses on my nose	5	3		1	4		6	7	
3: It fits perfectly on my nose	23	19		24	23		47	42	

Table A5. *Cont.*

Question	SV Group (n = 30)			PAL Group (n = 30)			All spectacles (n = 60)		
	Personalized	Control	p	Personalized	Control	p	Personalized	Control	p
How are the temples? (1–10)	8.53 ± 1.66	8. 20 ± 2.19	0.58	8.83 ± 1.08	8.53 ± 1.33	.27	8.68 ± 1.40	8.37 ± 1.80	.31
How do you find the temples according to your head?									
1: They are too loose on my head	8	9	0.96	6	9	0.33	14	18	0.38
2: They press on my head	1	0		0	3		1	3	
3: They fit perfectly according to my head	21	21		24	18		45	39	
How is the length of the temples?									
1: Too long	7	11	0.52	4	6	0.69	11	17	0.29
2: Too short	1	0		2	1		3	1	
3: The temple's length is exactly what I need	22	19		24	23		46	42	
How is your vision in general with the new spectacles? (1–10)	9.77 ± 0.43	9.37 ± 1.03	0.05	8.47 ± 1.33	7.90 ± 2.41	0.51	9.12 ± 1.18	8.63 ± 1.98	0.19
How do you find the field of view with the new spectacles in general? (1–10)	8. 40 ± 1.96	8.30 ± 1.66	0.59	8.70 ± 1.34	8.40 ± 1.92	0.52	8.55 ± 1.67	8.35 ± 1.78	0.34
Do you notice any distortion of images or blurring?									
1: No, not at all	27	24	0.28	23	20	0.39	50	44	0.18
2: Slightly	3	6		7	10		10	16	
Do you experience any swaying feelings?									
1: No, not at all	24	22	0.54	17	19	0.59	41	41	0.99
2: Slightly	6	8		13	11		19	19	
OVERALL SCORE	43.63 ± 4.73	40.63 ± 8.32	0.20	42.90 ± 4.37	42.13 ± 5.30	0.95	43.27 ± 4.52	41.38 ± 6.96	0.30

Note: Patient satisfaction was assessed linearly on a 1–10-point scale, with 10 indicating the best quality of vision or comfort and 1 indicating the worst. SV: single vision; PAL: progressive addition lens; p: Wilcoxon signed-rank test and Chi-squared test for the 2 × 2 or 3 × 2 contingency tables. Data: mean ± standard deviation.

Table A6. Scores of the self-evaluation questionnaires in the single vision (SV) and progressive addition lenses (PAL) groups.

Question	SV Group (n = 30)			PAL Group (n = 30)			All spectacles (n = 60)		
	Personalized	Control	p	Personalized	Control	p	Personalized	Control	p
How comfortable is this pair of spectacles in general? (1–5)	4.13 ± 1.01	3.73 ± 1.26	0.17	4.13 ± 0.94	3.90 ± 1.06	0.32	4.13 ± 0.96	3.82 ± 1.16	0.09
How easily and quickly could you adapt to this pair of spectacles? (1–5)	4.13 ± 1.01	4.0 ± 1.44	0.76	3.80 ± 1.16	3.63 ± 1.43	0.59	3.97 ± 1.09	3.82 ± 1.43	0.55
Spectacles slide down from the nose (1–5)	3.57 ± 1.22	3.33 ± 1.40	0.51	3.80 ± 1.27	3.93 ± 1.23	0.56	3.68 ± 1.24	3.63 ± 1.34	0.89
Lenses steam up while wearing them (1–5)	4.33 ± 1.03	4.20 ± 1.09	0.58	4.23 ± 0.90	4.53 ± 0.63	0.06	4.28 ± 0.96	4.37 ± 0.90	0.51

Table A6. *Cont.*

Question	SV Group (n = 30)			PAL Group (n = 30)			All spectacles (n = 60)		
	Personalized	Control	p	Personalized	Control	p	Personalized	Control	p
How comfortable was this pair on your nose during the two weeks of wearing them? (1–5)	3.70 ± 1.26	3.53 ± 1.55	0.61	4.23 ± 0.77	4.07 ± 1.05	0.33	3.97 ± 1.07	3.80 ± 1.34	0.34
How comfortable were the temples during the two weeks of wearing this pair? (1–5)	4.30 ± 1.02	3.77 ± 1.48	0.19	3.99 ± 0.99	4.17 ± 0.95	0.38	4.13 ± 1.02	3.97 ± 1.25	0.46
How satisfied are you with your vision/visual acuity using this pair looking:									
At far distances (>4–6 m: TV. . .)? (1–5)	4. 87 ± 0.34	4.70 ± 0.70	0.29	4.43 ± 0.86	3.80 ± 1.21	0.02	4.65 ± 0.68	4.25 ± 1.08	0.01
At intermediate distances (≈60 cm–1 m: computer work) (1–5)	4.83 ± 0.38	4.73 ± 0.58	0.40	4.03 ± 1.22	3.97 ± 1.25	0.96	4.43 ± 0.98	4.35 ± 1.04	0.77
Doing your near work (≈30–50 cm: reading...) (1–5)	4.77 ± 0.73	4.67 ± 0.61	0.35	4.20 ± 1.21	3.60 ± 1.35	0.02	4.48 ± 0.95	4.13 ± 1.17	0.01
How satisfied are you with the visual field using this pair:									
At far distances (>4–6 m: TV. . .)? (1–5)	4.03 ± 1.07	4.27 ± 1.17	0.16	4.37 ± 0.99	4.00 ± 1.20	0.10	4.20 ± 1.04	4.13 ± 1.17	0.94
At intermediate distances (≈60 cm–1 m: computer work) (1–5)	4.27 ± 0.91	4.47 ± 0.97	0.19	4.20 ± 1.06	4.07 ± 1.20	0.38	4.23 ± 0.98	4.27 ± 1.10	0.75
Doing your near work (≈30–50 cm: reading...) (1–5)	4.57 ± 0.73	4.57 ± 0.94	0.72	4.27 ± 0.94	4.00 ± 1.26	0.19	4.42 ± 0.85	4.28 ± 1.14	0.41
Did you experience any swaying feeling when using this pair? (1–5)	4. 50 ± 0.73	4.20 ± 1.18	0.19	4.30 ± 1.09	3.80 ± 1.15	0.03	4.40 ± 0.92	4.00 ± 1.18	0.02
How satisfied are you with this eyewear in general? (1–5)	4.10 ± 0.96	3.77 ± 1.22	0.25	4.17 ± 0.87	3.70 ± 1.09	0.04	4.13 ± 0.91	3.73 ± 1.15	0.03
OVERALL SCORE	60.10 ± 7.30	57.93 ± 10.50	0.48	58.13 ± 9.25	55.17 ± 11.18	0.21	59.12 ± 8.32	56.55 ± 10.87	0.18

Note: Patient satisfaction was assessed linearly on a 1–5-point scale, with 5 indicating the best quality of vision or comfort and 1 indicating the worst. SV: single vision; PAL: progressive addition lens p: Wilcoxon signed-rank test. Data: mean ± standard deviation.

References

1. Sheppard, A.L.; Wolffsohn, J.S. Digital eye strain: Prevalence, measurement and amelioration. *BMJ Open Ophthalmol.* **2018**, *3*, e000146. [CrossRef]
2. Wolffsohn, J.S.; Davies, L.N. Presbyopia: Effectiveness of correction strategies. *Prog. Retin. Eye Res.* **2019**, *68*, 124–143. [CrossRef]
3. Jaschinski, W.; König, M.; Mekontso, T.M.; Ohlendorf, A.; Welscher, M. Comparison of progressive addition lenses for general purpose and for computer vision: An office field study. *Clin. Exp. Optom.* **2015**, *98*, 234–243. [CrossRef] [PubMed]
4. De Lestrange-Anginieur, E.; Kee, C.S. Optical performance of progressive addition lenses (PALs) with astigmatic prescription. *Sci. Rep.* **2021**, *11*, 2984. [CrossRef] [PubMed]
5. Kee, C.-S.; Leung, T.W.; Kan, K.-H.B.; Lam, C.H.-I.B. Effects of Progressive Addition Lens Wear on Digital Work in Pre-presbyopes. *Optom. Vis. Sci.* **2018**, *95*, 457–467. [CrossRef] [PubMed]
6. Meister, D.J.; Fisher, S.W. Progress in the spectacle correction of presbyopia. Part 1: Design and development of progressive lenses. *Clin. Exp. Optom.* **2008**, *91*, 240–250. [CrossRef]
7. Meister, D.J.; Fisher, S.W. Progress in the spectacle correction of presbyopia. Part 2: Modern progressive lens technologies. *Clin. Exp. Optom.* **2008**, *91*, 251–264. [CrossRef]

8. Sheedy, J.E. Progressive addition lenses—Matching the specific lens to patient needs. *Optom. J. Am. Optom. Assoc.* **2004**, *75*, 83–102. [CrossRef]
9. Han, Y.; Ciuffreda, K.J.; Selenow, A.; Ali, S.R. Dynamic interactions of eye and head movements when reading with single-vision and progressive lenses in a simulated computer-based environment. *Investig. Opthalmology Vis. Sci.* **2003**, *44*, 1534–1545. [CrossRef]
10. Legras, R.; Vincent, M.; Marin, G. Does visual acuity predict visual preference in progressive addition lenses? *J. Optom.* **2023**, *16*, 91–99. [CrossRef]
11. Sullivan, C.M.; Fowler, C.W. Investigation of progressive addition lens patient tolerance to dispensing anomalies. *Ophthalmic Physiol. Opt.* **1990**, *10*, 16–20. [CrossRef] [PubMed]
12. Sheedy, J.; Hardy, R.F.; Hayes, J.R. Progressive addition lenses—Measurements and ratings. *Optom. J. Am. Optom. Assoc.* **2006**, *77*, 23–39. [CrossRef]
13. Huang, C.-Y.; Raasch, T.W.; Yi, A.Y.; Bullimore, M.A. Comparison of Progressive Addition Lenses by Direct Measurement of Surface Shape. *Optom. Vis. Sci.* **2013**, *90*, 565–575. [CrossRef]
14. Barbero, S.; Portilla, J. The relationship between dioptric power and magnification in progressive addition lenses. *Ophthalmic Physiol. Opt.* **2016**, *36*, 421–427. [CrossRef] [PubMed]
15. Rifai, K.; Wahl, S. Specific eye–head coordination enhances vision in progressive lens wearers. *J. Vis.* **2006**, *16*, 5. [CrossRef]
16. Hutchings, N.; Irving, E.L.; Jung, N.; Dowling, L.M.; Wells, K.A. Eye and head movement alterations in naïve progressive addition lens wearers. *Ophthalmic Physiol. Opt.* **2007**, *27*, 142–153. [CrossRef]
17. Alvarez, T.L.; Kim, E.H.; Granger-Donetti, B. Adaptation to Progressive Additive Lenses: Potential Factors to Consider. *Sci. Rep.* **2017**, *7*, 2529. [CrossRef]
18. Pascual, E.; Gómez-Pedrero, J.A.; Alonso, J. Theoretical performance of progressive addition lenses with poorly measured individual parameters. *Ophthalmic Physiol. Opt.* **2023**, *43*, 244–253. [CrossRef]
19. Han, S.C.; Graham, A.D.; Lin, M.C. Clinical Assessment of a Customized Free-Form Progressive Add Lens Spectacle. *Optom. Vis. Sci.* **2011**, *88*, 234–243. [CrossRef]
20. Hoya Corporation. White Paper Hoyalux iD MyStyle V+ 2014, 01. Available online: https://www.hoyavision.com/contentassets/cd44fd98c9fb469497d6621fe1db16d9/id-mystyle2-whitepaper_full-version_10_21_20.pdf/ (accessed on 5 August 2023).
21. Radner, W.; Obermayer, W.; Richter-Mueksch, S.; Willinger, U.; Velikay-Parel, M.; Eisenwort, B. The validity and reliability of short German sentences for measuring reading speed. *Graefe's Arch. Clin. Exp. Ophthalmol.* **2002**, *240*, 461–467. [CrossRef]
22. Stifter, E.; König, F.; Lang, T.; Bauer, P.; Richter-Müksch, S.; Velikay-Parel, M.; Radner, W. Reliability of a standardized reading chart system: Variance component analysis, test-retest and inter-chart reliability. *Graefe's Arch. Clin. Exp. Ophthalmol.* **2004**, *242*, 31–39. [CrossRef] [PubMed]
23. Selenow, A.; Bauer, E.A.; Ali, S.R.; Spencer, L.W.; Ciuffreda, K.J. Assessing Visual Performance with Progressive Addition Lenses. *Optom. Vis. Sci.* **2002**, *79*, 502–505. [CrossRef]
24. Sánchez-Brau, M.; Domenech-Amigot, B.; Brocal-Fernández, F.; Quesada-Rico, J.A.; Seguí-Crespo, M. Prevalence of Computer Vision Syndrome and Its Relationship with Ergonomic and Individual Factors in Presbyopic VDT Workers Using Progressive Addition Lenses. *Int. J. Environ. Res. Public Health* **2020**, *17*, 1003. [CrossRef] [PubMed]
25. Garcia-Espinilla, O.; Gallegos-Cocho, I.; Sanchez, I.; Cañadas, P.; Martin, R. Comparison of physiognomy and frame angle parameters using different devices to prescribe progressive addition lenses. *Clin. Exp. Optom.* **2022**, *105*, 420–427. [CrossRef] [PubMed]
26. Garcia-Espinilla, O.; Gallegos-Cocho, I.; Sanchez, I.; Cañadas, P.; Martin, R. Interdevice agreement in the measurement of physiognomy parameters and frame angles to prescribe progressive addition lenses. *Clin. Exp. Optom.* **2023**, *106*, 69–74. [CrossRef]
27. Fontaine, N.O.; Hanssens, J.-M.O.; Nguyen, M.; Bérubé, O. Ordering Eyeglasses Using 3D Head Scan Technology versus Established Online and Storefront Clinic Methods. *Optom. Vis. Sci.* **2023**, *100*, 319–327. [CrossRef]
28. Alionte, C.G.; Ungureanu, L.M.; Alexandru, T.M. Innovation Process for Optical Face Scanner Used to Customize 3D Printed Spectacles. *Materials* **2022**, *15*, 3496. [CrossRef]
29. Campomanes, A.G.d.A.; Meer, E.; Clarke, M.; Brodie, F.L. Using a Smartphone 3-Dimensional Surface Imaging Technique to Manufacture Custom 3-Dimensional–Printed Eyeglasses. *JAMA Ophthalmol* **2022**, *140*, 966–973. [CrossRef]
30. Lee, L.; Burnett, A.M.; Panos, J.G.; Paudel, P.; Keys, D.; Ansari, H.M.; Yu, M. 3-D printed spectacles: Potential, challenges and the future. *Clin. Exp. Optom.* **2020**, *103*, 590–596. [CrossRef]

life

MDPI

Article

The Relationship between Fixation Stability and Retinal Structural Parameters in Children with Anisometropic, Strabismic and Mixed Amblyopia

Raquel Mompart-Martínez [1,2], Marc Argilés [2,3], Genis Cardona [2,4,*], Lluís Cavero-Roig [1], Lluís González-Sanchís [1] and Maria Soledad Pighin [1]

1 Institut Català de la Retina (ICR), 08022 Barcelona, Spain; mompart.r@gmail.com (R.M.-M.); lluis.cavero@icrcat.com (L.C.-R.); lgonzalez@icrcat.com (L.G.-S.); solepin@gmail.com (M.S.P.)
2 Department of Optics and Optometry, Universitat Politècnica de Catalunya (UPC), 08222 Terrassa, Spain; marc.argiles@upc.edu
3 Center for Sensors, Instruments, and Systems Development (CD6), Universitat Politècnica de Catalunya (UPC), 08222 Terrassa, Spain
4 Applied Optics and Image Processing Group (GOAPI), Universitat Politècnica de Catalunya (UPC), 08222 Terrassa, Spain
* Correspondence: genis.cardona@upc.edu

Abstract: (1) Background: Amblyopia is an ocular condition leading to structural and functional changes. The relationship between these changes is complex and remains poorly understood. (2) Methods: Participants included 31 children aged 5 to 9 years with strabismic ($n = 9$), anisometropic ($n = 16$) and mixed ($n = 6$) unilateral amblyopia, and 14 age-matched non-amblyopic children. The 95% and 63% Bivariate Contour Ellipse Area (BCEA), axial length, Foveal Avascular Zone (FAZ) area, center macular thickness and volume were assessed. The relationship between these parameters was explored. (3) Results: Statistically significant differences were found among the four groups in best corrected distance visual acuity (BCVA) ($p < 0.001$), BCEA 95% ($p = 0.002$) and BCEA 63% ($p = 0.002$), but not in the FAZ area, central macular thickness, central macular volume and axial length. Eyes with amblyopia had poorer BCVA and larger fixation instability than controls. Inter-ocular differences were more significant in patients with strabismic amblyopia, particularly in BCVA ($p = 0.003$), central macular thickness ($p < 0.001$) and central macular volume ($p = 0.002$). In amblyopic eyes, BCEA 95% and 63% were correlated with BCVA, but not with the FAZ area. (4) Conclusion: Amblyopia is associated with a reduction in fixation stability and BCVA, although there is a general lack of correlation with structural changes, suggesting a complex interaction between anatomy and function in amblyopia.

Keywords: amblyopia; fixation stability; retinal microvasculature; macular thickness; macular volume; stereoacuity; strabismus; anisometropia

Citation: Mompart-Martínez, R.; Argilés, M.; Cardona, G.; Cavero-Roig, L.; González-Sanchís, L.; Pighin, M.S. The Relationship between Fixation Stability and Retinal Structural Parameters in Children with Anisometropic, Strabismic and Mixed Amblyopia. *Life* **2023**, *13*, 1517. https://doi.org/10.3390/life13071517

Academic Editors: José-María Sánchez-González, Carlos Rocha-de-Lossada and Alejandro Cerviño

Received: 13 June 2023
Revised: 28 June 2023
Accepted: 5 July 2023
Published: 6 July 2023

1. Introduction

Amblyopia is a neurodevelopmental disorder that results from an abnormal visual experience during a critical period of visual development [1,2]. Risk factors that typically contribute to and serve to classify amblyopia include strabismus (strabismic amblyopia), anisometropic uncorrected refractive error (anisometric amblyopia) or a combination of both (mixed amblyopia) [2,3]. Several studies have reported possible structural changes in the visual cortex [4] and lateral geniculate nucleus [5] in amblyopic eyes, compared with normal eyes [6], as well as differences in macular thickness and volume [7,8], retinal nerve fiber layer (RNFL) [8,9], and macular capillary vascular structure, including the foveal avascular zone (FAZ) area [10–14].

Current developments in microperimetry have allowed researchers to explore fixation stability in amblyopia [1,15–21] and the relationship between fixation stability, visual acuity and stereoacuity in different types of amblyopia, which remains poorly understood [15,16]. The analysis of fixation stability has also been employed to assess the effectiveness of different types of treatments for amblyopia, such as part-time occlusion of the non-amblyopic eye [22–24] or surgical intervention to correct strabismus [25].

Previous researchers have explored the association between retinal structural parameters, such as the FAZ area, and visual function parameters, including visual acuity, visual fields and fixation stability, in retinal vein occlusion, diabetic retinopathy and other pathologies [26–28]. However, the relationship between fixation stability and specific retinal anatomical features in various types of amblyopia is not conclusive. Similarly, several studies have investigated differences in the FAZ area between amblyopic and contralateral eyes, as well as healthy controls, reporting contradictory evidence [10,12,29–32]. These discrepancies may be due to variations in sample demographics, such as age (adults or children) and distribution of amblyopia types, as well as differences in instrumentation, analysis software and methodology. Although this previous research has shown vascular anomalies in amblyopic patients in terms of density and structure [10,12], less attention has been paid to the possible relationship between fixation stability, as determined by the bivariate contour ellipse area (BCEA) and the FAZ area. Indeed, although one study failed to find any correlation between these parameters in patients with retinopathy and prematurity [28], to the best of our knowledge, it has not been studied in amblyopia and different subtypes of strabismus. As previous researchers have observed different patterns of visual function depending on the degree and type of amblyopia [33], it may be relevant to explore fixation stability and structural parameters in different subtypes of amblyopia and strabismus.

Thus, it was the aim of the present study to assess the possible correlation between fixation stability, defined by BCEA, and structural parameters, such as the FAZ area, in a sample of children with unilateral strabismic, anisometropic and mixed amblyopia. Additional study variables were axial length (AL), best corrected distance visual acuity (BCVA) and stereoacuity, as well as central macular thickness and volume. Results from the amblyopic eyes were compared with those of the contralateral eye, and with those of a non-amblyopic control group.

2. Materials and Methods

2.1. Study Sample

A cross-sectional study was conducted at the Institut Català de la Retina (ICR, Barcelona, Spain) from September 2021 to May 2022. The study was approved by the Ethics Committee of the Grupo Hospitalario Quirón Salud, (ICR-13/21-2022/10-OFT-ICR) and was compliant with the principles of the Declaration of Helsinki. Informed consent was obtained from legal guardians after a detailed explanation of the nature and possible consequences of the study.

Children aged 5 to 9 years were recruited for this investigation. Participants in the study had either strabismic, anisometropic or mixed unilateral amblyopia. Amblyopia was defined as BCVA of $\geq$0.2 logMAR for the worse eye and an interocular difference of $\geq$0.15 logMAR. In addition, for anisometropic amblyopia, interocular differences in refractive error needed to be of 2.00 diopters (D) or more in sphere and/or 1.50 D in cylinder. No anisohyperopic participants were included in the study. In addition, strabismic amblyopia required a minimum angle of manifest deviation of 5 prism diopters. Patients diagnosed with neurological pathologies, retinopathies and/or maculopathies, glaucoma, nystagmus, media opacities, systemic diseases, cardiovascular or renal diseases, retardation and/or prematurity were excluded from the study, as were those with a manifest deviation of 30 prism diopters or more. In addition, a control age and gender-matched group of healthy non-amblyopic subjects was included.

2.2. Procedure

Monocular BCVA was evaluated with the Early Treatment Diabetic Retinopathy Study (ETDRS) test at 4 m and recorded letter by letter in logMAR notation. Stereoacuity was evaluated with the TNO stereo test, with the aid of red–green filters (a score of 800 arc seconds was assigned to patients failing to identify any of the pictures). For both BCVA and stereoacuity, patients received instructions on the procedure and an initial trial was conducted to ensure they fully understood what was expected of them. For this trial, the largest letter of the ETDRS test was used for BCVA measurements, and the butterfly plate of the TNO test for stereoacuity, which corresponds to 1500 arc seconds. Exactly the same procedure was employed for all patients and all measurements were conducted by an examiner unaware of the patient group allocation, although in patients with strabismic amblyopia this single-blind experimental condition could not be maintained. All patients included in the study were able to follow the BCVA and stereoacuity measurement procedures without difficulties.

Refractive error was measured using the retinoscopy technique, with and without cycloplegia. Visual alignment was measured with the cover test in both near and far vision, and the angle of deviation of strabismic patients was determined with the cover test and a prismatic bar, both in conditions of best corrected visual acuity.

Fixation stability was measured using the Macular Analyzer Integrity Assessment (MAIA) microperimeter (CenterVue, Padova, Italy) and followed the BCEA methodology described by Crossland and co-workers, with a fixation strategy of 30 s [34]. Briefly, for each eye, assessment of fixation stability begins with the capture of a reference fundus image and the identification by the examiner of a high contrast macular retinal landmark. During the subsequent 30 s test period, the patient is instructed to look at the fixation stimulus (red dot) while the MAIA software (version 2.6.0) determines the shift between the reference image and the real-time fundus image at 25 Hz, obtaining 750 sets of coordinates (X and Y) to describe fixation changes. The software then defines the BCEA as the best fit elliptic contour containing either 95% (BCEA 95%) or 63% (BCEA 63%) of the fixation points, with smaller BCEA values denoting better fixation stability (see, for example, Figure 1). The MAIA automatically corrects for refraction errors in the spherical equivalent range of −15.00 to +10.00 D, allowing for the measurements to be conducted without habitual refractive correction.

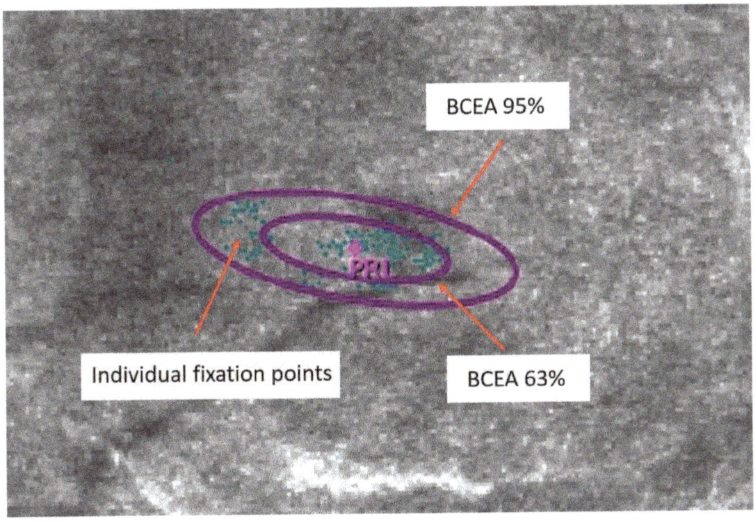

Figure 1. Best fit elliptic contour containing either 95% (BCEA 95%) or 63% (BCEA 63%) of the fixation points in a patient with strabismic amblyopia. PRL: Preferred Retinal Locus.

The FAZ area of the superficial capillary plexus (SCP) was measured with the OCT-A Spectralis (Heidelberg Engineering, Heidelberg, Germany) with angiography software. To determine the FAZ area, a scan pattern of 10 × 10 degrees (consisting of 512 scanned sections separated by 6 μm) was centered on the fovea. Internal fixation was used to ensure proper alignment of the eye. It may be noted that in certain patients with large fixation instability, internal fixation may be eccentric instead of central. However, if the fixation moved outside of the initial internal fixation zone, the OCT stopped the test until the patient correctly fixed the stimulus again. Each scan was automatically segmented. Finally, the FAZ area was calculated manually employing the tools provided by the Spectralis software: the boundaries of the FAZ area were delineated by the freehand tool and calculated in mm^2, as previously described (Figure 2) [35]. Axial length (AL) was measured with the ZEISS IOLMaster 700 biometer (Carl Zeiss Meditec, Jena, Germany).

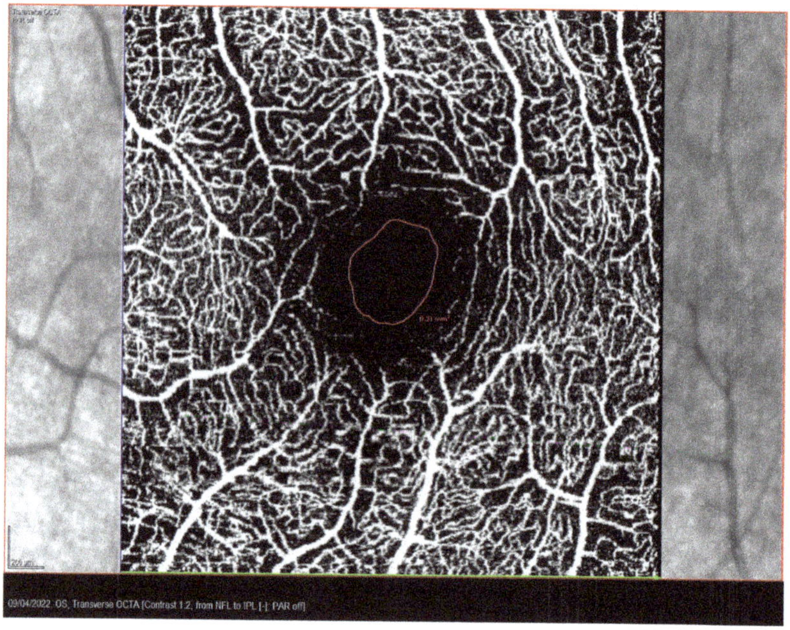

Figure 2. Manually delimited foveal vascular zone (FAZ) area in a patient with strabismic amblyopia.

Macular thickness and volume were measured using the OCT-A Spectralis. A macular cube scan pattern of 20 × 15 degrees (consisting of 512 HR A-scans, 37 sections separated by 120 μm) centered on the fovea was acquired. Internal fixation was used for proper alignment of the eye. Each scan was automatically segmented and an optometrist examined all macular cubes to ensure that the analyzed macular thickness and volume corresponded to the foveal center.

2.3. Data Analysis

SPSS version 27 (IBM Corp., Armonk, NY, USA) for Windows was used for data analysis. Values of BCEA were transformed to their corresponding $\log_{10}$ values. The Shapiro–Wilk test was employed to explore data normality, whereupon results are accordingly described as mean and standard deviation (SD), or median and range. Inferential analysis was conducted with the ANOVA or Kruskal–Wallis tests, with the corresponding post-hoc Bonferroni or Dunn–Bonferroni pair-wise analysis and correction to account for multiple comparisons, or with the paired Student's *t*-test or Wilcoxon tests. In addition, possible associations between variables were analyzed with the Spearman coefficient of correlation test, and the Chi-Squared test was employed for frequency distribution analysis. A p-value less than 0.05 was considered to denote statistical significance.

Sample size calculation was performed using G*Power version 3.1.9.2 software (Heinrich-Heine-University Dusseldorf, Dusseldorf, Germany). The study of Subramanian and co-workers [15] was used as a reference for sample size calculation. The main outcome for this calculation was BCEA, with an estimated common SD of 0.38 log deg^2 and a minimum expected difference between amblyopic and fellow eyes of 0.36 log deg^2, and considering a 95% statistical power. The required minimum sample size was 18 participants per group for pair-wise comparisons.

3. Results

A total of 31 children were included in the amblyopic group (15 boys, 14 girls) aged 5 to 9 years (mean ± SD of 6.6 ± 1.1 years), with strabismic (n = 9), anisometropic (n = 16) and mixed (n = 6) amblyopia. The control group included 14 non-amblyopic children (7 boys, 7 girls), aged 6 to 9 years (7.6 ± 1.0 years). No statistically significant difference was found in neither age nor sex distribution between the three amblyopic and control groups ($p > 0.05$).

For the analysis, first eyes with different types of amblyopia were compared amongst them and with one of the eyes of subjects from the control group. Table 1 presents a summary of the results of monocular BCVA, stereoacuity, log BCEA 95%, log BCEA 63%, FAZ area, central macular thickness, central macular volume and axial length of strabismic, anisometropic and mixed amblyopic eyes. To compare with amblyopic eyes, the non-dominant eye was selected from the control group, except for stereoacuity measurements. Statistically significant differences were found among the four groups in BCVA ($p < 0.001$), stereoacuity ($p < 0.001$), BCEA 95% ($p = 0.002$) and BCEA 63% ($p = 0.002$), but not in the FAZ area, central macular thickness, central macular volume and axial length (all $p > 0.05$). A pair-wise analysis with the post-hoc Dunn–Bonferroni test revealed statistically significant differences in BCVA and stereoacuity between control and strabismic, anisometric and mixed amblyopic groups (all $p < 0.001$), with superior values of both visual function parameters in the control group. Regarding BCEA 95% and BCEA 63%, the Dunn–Bonferroni or Bonferroni tests revealed statistically significant differences between control eyes and strabismic and anisometropic eyes ($p = 0.001$ and $p = 0.034$, respectively, for BCEA 95%; $p = 0.004$ and $p = 0.018$, respectively, for BCEA 63%), but not between mixed amblyopic and control eyes. Strabismic eyes tended to have the largest BCEA 95% and BCEA 63% results. By pooling the data from all amblyopic groups, statistically significant differences were found between BCEA 95% and BCEA 63% ($p < 0.001$).

Table 1. Best corrected distance visual acuity (BCVA), stereoacuity, log BCEA 95%, log BCEA 63%, FAZ area, central macular thickness, central macular volume and axial length for strabismic, anisometropic and mixed amblyopic eyes and control eyes (one eye per control subject was selected, except for stereoacuity evaluation). Results are shown as either mean ± SD or median (range) according to the normality of their distribution. The outcome of the ANOVA or Kruskal–Wallis tests (according to data normality) is displayed in the rightmost column.

	Strabismic Amblyopia	Anisometropic Amblyopia	Mixed Amblyopia	Control	p-Value
BCVA (logMAR)	0.362 ± 0.267	0.250 (0.800)	0.450 ± 0.356	0.000 (0.100)	<0.001
Stereoacuity (arc seconds)	400 (700)	200 (720)	800 (400)	40 (15)	<0.001
Log BCEA 95% (log deg^2)	0.323 ± 0.496	0.079 ± 0.354	0.110 ± 0.425	−0.316 ± 0.255	0.002
Log BCEA 63% (log deg^2)	−0.143 ± −0.496	−0.391 ± 0.336	−0.397 ± 0.467	−0.699 (0.602)	0.006
FAZ area (mm^2)	0.270 ± 0.173	0.369 ± 0.190	0.354 ± 0.113	0.269 ± 0.112	0.277
Central macular thickness (µm)	269.0 (83.0)	269.4 ± 20.0	251.5 ± 23.7	261.1 ± 27.6	0.496
Central macular volume (μm^3)	0.21 ± 0.01	0.22 ± 0.01	0.21 ± 0.01	0.22 (0.37)	0.291
Axial length (mm)	21.86 ± 0.62	22.34 ± 1.79	22.22 (6.41)	22.47 ± 1.20	0.563

For the second part of the analysis, inter-ocular differences between the amblyopic eye (or non-dominant eye in controls) and the dominant eyes were calculated for amblyopic patients and control subjects. Table 2 presents a summary of the inter-ocular differences in functional and structural parameters. Statistically significant inter-ocular differences were predominant in patients with anisometropic amblyopia (BCVA, $p = 0.003$; central macular thickness, $p < 0.001$; central macular volume, $p = 0.002$). No statistically significant interocular difference was found in control subjects.

Table 2. Inter-ocular differences in best corrected distance visual acuity (BCVA), log BCEA 95%, log BCEA 63%, FAZ area, central macular thickness, central macular volume and axial length for amblyopic and control groups (determined as amblyopic eye minus dominant eye in the amblyopic groups, or non-dominant eye minus dominant eye in the control group). Results are shown as either mean ± SD or median (range) according to the normality of their distribution, together with the outcome of the paired Student's *t*-test or the Wilcoxon test (according to data normality).

	Strabismic Amblyopia	Anisometropic Amblyopia	Mixed Amblyopia	Control
BCVA (logMAR)	0.189 ± 0.226 $p = 0.050$	0.194 ± 0.191 $p = 0.003$	0.333 ± 0.367 $p = 0.106$	−0.007 (0.100) $p = 1.000$
Log BCEA 95% (log deg^2)	0.235 ± 0.380 $p = 0.097$	0.240 ± 0.433 $p = 0.066$	0.003 ± 0.449 $p = 0.855$	−0.052 (0.862) $p = 0.132$
Log BCEA 63% (log deg^2)	0.288 ± 0.421 $p = 0.058$	0.236 ± 0.424 $p = 0.086$	−0.069 ± 0.428 $p = 1.000$	−0.038 (0.778) $p = 0.418$
FAZ area (mm^2)	−0.025 (1.060) $p = 0.310$	0.033 ± 0.091 $p = 0.105$	−0.002 ± 0.081 $p = 1.000$	−0.007 (0.130) $p = 0.571$
Central macular thickness (μm)	16.7 ± 17.2 $p = 0.025$	19.1 ± 14.8 $p < 0.001$	21.0 (19.0) $p = 0.063$	−0.143 (22.0) $p = 1.000$
Central macular volume (μm^3)	0.01 ± 0.01 $p = 0.073$	0.01 (0.03) $p = 0.002$	0.01 ± 0.01 $p = 0.181$	0.00 (0.37) $p = 0.098$
Axial length (mm)	−0.12 ± 0.19 $p = 0.074$	−0.26 (6.5) $p = 0.438$	0.64 ± 1.5 $p = 1.000$	0.00 (0.33) $p = 0.483$

Inter-ocular differences amongst groups in BCVA, BCEA 95%, BCEA 63% and central macular thickness were statistically significant ($p = 0.007$, $p = 0.043$, $p = 0.026$ and $p < 0.001$, respectively). Post-hoc pair-wise analysis revealed statistically significant inter-ocular differences in BCVA between the anisometropic and control groups ($p = 0.007$) and between the mixed and control groups ($p = 0.018$); in BCEA 95% between anisometropic and control groups ($p = 0.033$); in BCEA 63% between the strabismic and control groups ($p = 0.042$), and between the anisometropic and control groups ($p = 0.045$); and in central macular thickness between all three amblyopic groups and the control group (strabismic, $p = 0.013$; anisometropic, $p < 0.001$; mixed, $p = 0.020$).

By pooling the data from all amblyopic groups, and considering only the amblyopic eyes, statistically significant moderate correlations were found between BCVA and both BCEA 95% ($\rho = 0.575$, $p < 0.001$) and BCEA 63% ($\rho = 0.580$, $p < 0.001$). BCEA 95% and 63% showed a very strong correlation ($\rho = 0.987$, $p < 0.001$). Statistically significant moderate to strong correlations were found in the non-dominant eyes of the control group between the FAZ area and both central macular thickness ($\rho = -0.562$, $p < 0.001$) and central macular volume ($\rho = -0.740$, $p < 0.001$). However, in the amblyopic groups, only a weak correlation was evidenced between the FAZ area and central macular thickness ($\rho = -0.378$, $p = 0.012$). Upon examining each amblyopic group independently, the only significant correlation was between BCVA and central macular thickness in the strabismic group ($\rho = -0.857$, $p = 0.024$), as well as between both BCEA outcomes. No statistically significant correlation was evidenced in any group between BCEA 95% or BCEA 63% and the FAZ area. In particular, in strabismic and mixed amblyopia, there was no correlation between angle of deviation and fixation stability. In addition, inter-ocular differences in BVCA in amblyopic eyes were weakly correlated with inter-ocular differences in BCEA 95% ($\rho = 0.313$, $p = 0.039$)

and BCEA 63% ($\rho = 0.321$, $p = 0.036$) and moderately correlated with inter-ocular differences in central macular thickness ($\rho = 0.447$, $p = 0.003$).

4. Discussion

The aim of the present study was to explore BCVA, stereoacuity, fixation stability and retinal structural parameters (FAZ area, central macular thickness and volume), as well as AL, in children with different types of amblyopia and in normal controls. The relationship between structural and functional parameters was investigated.

Fixation stability, determined with the BCEA 95% and 63% was found to be particularly compromised in strabismic and, to a less extend, in anisometropic amblyopia, as compared with control eyes. No difference between control eyes and those with mixed amblyopia was found, although the reduced sample size for this type of amblyopia led to a slightly underpowered analysis, thus increasing the probability of type I error. These findings are in agreement with previous reports [15,19]. For instance, with the research of Subramanian et al. (2013), who also documented a reduction in stereoacuity related to amblyopia [15], as described in the present study. These authors attributed the discrepancies in the BCEA values between their study and published literature to the actual BCEA percentage under analysis [15,19]. To our knowledge, no previous research has explored both BCEA 95% and BCEA 63% from the same sample of patients. Although BCEA 95% and BCEA 63% were found to present a very strong correlation, a statistically significant difference was found between their values, underlining the need to interpret data with caution when comparing studies using different BCEA percentages. In addition, discrepancies in BCEA values among studies may be accounted for by the differences in age of participants and instrumentation.

To determine fixation stability, a fixation strategy of 30 s was implemented, which is the current minimum available interval for the MAIA microperimeter to provide reliable measurements. Given the age of some of the participants, they needed encouragement to maintain fixation on the target stimulus (for instance, they were instructed to carefully check the red fixation stimulus and tell the examiner immediately if it changed color to green). Previous researchers have employed longer fixation strategies, of 45 s or even 1 min intervals [36]. It must be highlighted that the software of MAIA constantly compares the fixation of the patient with the initial reference point defined by the examiner and automatically stops the measurement if fixation is lost or if the patient performs intrusive saccades, renewing measurements once the patient has gained the initial fixation reference. Although differences in instrumentation and sample characteristics may advise against any direct comparison, the variance in BCEA values of the present sample of young children is comparable to that obtained in previous research in adults, which may reflect that measurements in children are no less reliable than in adults, given the necessary precautions are taken [37,38].

No statistically significant differences were found in the FAZ area between amblyopic and control eyes, albeit a trend was observed in which anisometropic and mixed amblyopic eyes had larger FAZ areas than strabismic and control eyes. These findings would be in agreement with previous literature documenting larger FAZ areas in the SCP layer in anisometropic amblyopia [29,30], albeit other researchers have noted smaller FAZ areas in patients with anisometropic amblyopia, with or without strabismus [12], and in strabismic amblyopia [32].

Upon examining inter-ocular differences in functional and structural parameters, that is, amblyopic eye versus the contralateral eye, or both eyes of the control group, statistically significant differences were predominant in patients with anisometropic amblyopia, particularly in BCVA, central macular thickness and central macular volume. Inter-ocular differences in BCEA 95% and BCEA 63% were larger in strabismic and anisometropic patients than in the control group. Similarly, inter-ocular differences in central macular thickness were more noticeable in the three amblyopic groups than in the control group.

These findings suggest that the study of inter-ocular asymmetry in functional and anatomical parameters may be valuable.

Correlation analysis revealed several statistically significant, moderate correlations among some functional parameters, as well as among several anatomical parameters, but no correlation was disclosed between functional and anatomical parameters, with the exception of BCVA and central macular thickness in the strabismic group. Thus, for instance, whereas in amblyopia both BCEA values were moderately correlated with BCVA, BCEA and the FAZ area did not display any significant correlation, which may suggest that the FAZ area does not influence fixation stability, although further research is needed to confirm this hypothesis. Indeed, the lack of correlation between fixation stability and the FAZ area has been previously reported for other retinal pathologies [26–28], but, as far as we know, it has not been explored in amblyopia. Similarly, the FAZ area was not correlated with BCVA, in contrast with the report by Huang et al. (2021) [31], in which the authors observed an improvement in both visual acuity and FAZ area following 6 months of amblyopia treatment.

This study had some limitations, mainly a reduced sample size and slightly underpowered analysis, considering a required estimated sample size of 18 per group. Unfortunately, these patients are relatively young and sometimes found it difficult to cooperate with the measurements and could not be included in the study. In cases of wide-angle strabismus, certain tests could not be performed due to saccadic re-fixational eye movements that prevented correct captures for scans. In addition, the manual nature of some of these measurements may contribute to reducing the reproducibility of some of the findings. Thus, FAZ area measurements were performed manually, following the border of the vascular and the avascular textures shown in the OCT-A image, as previously described [35]. There are some OCT-A devices with integrated software for quantitative analysis, such as the OCT 5000 (Zeiss Meditec Inc., Dublin, CA, USA) or the RTVue XR Avanti (Optovue Inc., Fremont, CA, USA). When this possibility is not available, third-party software, such as the Fiji toolbox, could be used to assist in this analysis.

In addition, although the TNO test has been documented to be adequate to measure stereoacuity in children aged 3 to 6 years old [39], there are currently other objective methods to estimate stereoacuity which may be implemented in future studies. For instance, a promising avenue of research, which has recently proved its clinical applicability in adults and children, is the analysis of ocular-following responses through video-oculography [40,41]. Briefly, ocular-following responses exhibit a strong binocular summation, which is sensitive to the interocular correlation between stimuli presented under binocular conditions, mediated by disparity-sensitive cortical neurons. It is therefore assumed that in patients with impaired stereoacuity, such as in amblyopia, ocular-following responses are compromised.

In conclusion, children with amblyopia have more fixation instability compared to normal controls. Although BCEA 95% and 63% show a strong correlation, these values are not interchangeable. Fixation instability is associated with reduced visual acuity, but not with the FAZ area, suggesting a complex interaction between structural and functional changes in amblyopia, which supports further investigation.

Author Contributions: Conceptualization, R.M.-M., M.A., L.C.-R. and M.S.P.; methodology, R.M.-M., M.A., L.C.-R. and M.S.P.; validation, R.M.-M., M.A., L.C.-R., L.G.-S. and M.S.P.; formal analysis, R.M.-M., M.A. and G.C.; investigation, R.M.-M.; resources, L.G.-S.; data curation, R.M.-M., M.A. and G.C.; writing—original draft preparation, R.M.-M. and M.A.; writing—review and editing, R.M.-M., M.A. and G.C.; visualization, R.M.-M.; supervision, M.A., L.C.-R., M.S.P.; project administration, L.G.-S. and M.S.P. All authors have read and agreed to the published version of the manuscript.

Funding: This research received no external funding.

Institutional Review Board Statement: This study was conducted in accordance with the Declaration of Helsinki, and approved by the Ethics Committee of the Grupo Hospitalario Quirón Salud, (ICR-13/21-2022/10-OFT-ICR).

Informed Consent Statement: Informed consent was obtained from all subjects involved in the study.

Data Availability Statement: Data may be available upon reasonable request to the authors.

Conflicts of Interest: The authors declare no conflict of interest.

References

1. Birch, E.E. Amblyopia and binocular vision. *Prog. Retin. Eye Res.* **2013**, *33*, 67–84. [CrossRef] [PubMed]
2. Kates, M.M.; Beal, C.J. Amblyopia. *JAMA* **2021**, *325*, 408–408. [CrossRef]
3. Magramm, I. Amblyopia: Etiology, Detection, and Treatment. *Pediatr. Rev.* **1992**, *13*, 7–14. [CrossRef] [PubMed]
4. Levi, D.M.; Knill, D.C.; Bavelier, D. Stereopsis and amblyopia: A mini-review. *Vis. Res.* **2015**, *114*, 17–30. [CrossRef]
5. Sloper, J. The other side of amblyopia. *J. Am. Assoc. Pediatr. Ophthalmol. Strabismus* **2016**, *20*, 1.e1–1.e13. [CrossRef] [PubMed]
6. Gaier, E.D.; Gise, R.; Heidary, G. Imaging Amblyopia: Insights from Optical Coherence Tomography (OCT). *Semin. Ophthalmol.* **2019**, *34*, 303–311. [CrossRef]
7. Chen, W.; Xu, J.; Zhou, J.; Gu, Z.; Huang, S.; Li, H.; Qin, Z.; Yu, X. Thickness of retinal layers in the foveas of children with anisometropic amblyopia. *PLoS ONE* **2017**, *12*, e0174537. [CrossRef]
8. Andalib, D.; Javadzadeh, A.; Nabai, R.; Amizadeh, Y. Macular and Retinal Nerve Fiber Layer Thickness in Unilateral Anisometropic or Strabismic Amblyopia. *J. Pediatr. Ophthalmol. Strabismus* **2013**, *50*, 218–221. [CrossRef]
9. Al-Haddad, C.E.; Mollayess, G.M.E.L.; Cherfan, C.G.; Jaafar, D.F.; Bashshur, Z.F. Retinal nerve fibre layer and macular thickness in amblyopia as measured by spectral-domain optical coherence tomography. *Br. J. Ophthalmol.* **2011**, *95*, 1696–1699. [CrossRef]
10. Can, G.D. Quantitative analysis of macular and peripapillary microvasculature in adults with anisometropic amblyopia. *Int. Ophthalmol.* **2020**, *40*, 1765–1772. [CrossRef]
11. Karabulut, M.; Karabulut, S.; Sül, S.; Karalezli, A. Microvascular differences in amblyopic subgroups: An observational case–control study. *Eur. J. Ophthalmol.* **2021**, *32*, 1242–1251. [CrossRef] [PubMed]
12. Araki, S.; Miki, A.; Goto, K.; Yamashita, T.; Yoneda, T.; Haruishi, K.; Ieki, Y.; Kiryu, J.; Maehara, G.; Yaoeda, K. Foveal avascular zone and macular vessel density after correction for magnification error in unilateral amblyopia using optical coherence tomography angiography. *BMC Ophthalmol.* **2019**, *19*, 171. [CrossRef] [PubMed]
13. Linderman, R.E.; Cava, J.A.; Salmon, A.E.; Chui, T.Y.; Marmorstein, A.D.; Lujan, B.J.; Rosen, R.B.; Carroll, J. Visual Acuity and Foveal Structure in Eyes with Fragmented Foveal Avascular Zones. *Ophthalmol. Retin.* **2020**, *4*, 535–544. [CrossRef] [PubMed]
14. Liu, C.; Zhang, Y.; Gu, X.; Wei, P.; Zhu, D. Optical coherence tomographic angiography in children with anisometropic amblyopia. *BMC Ophthalmol.* **2022**, *22*, 269. [CrossRef]
15. Subramanian, V.; Jost, R.M.; Birch, E.E. A Quantitative Study of Fixation Stability in Amblyopia. *Investig. Opthalmol. Vis. Sci.* **2013**, *54*, 1998–2003. [CrossRef] [PubMed]
16. Kelly, K.R.; Cheng-Patel, C.S.; Jost, R.M.; Wang, Y.-Z.; Birch, E.E. Fixation instability during binocular viewing in anisometropic and strabismic children. *Exp. Eye Res.* **2019**, *183*, 29–37. [CrossRef]
17. Shaikh, A.G.; Otero-Millan, J.; Kumar, P.; Ghasia, F.F. Abnormal Fixational Eye Movements in Amblyopia. *PLoS ONE* **2016**, *11*, e0149953. [CrossRef]
18. Chung, S.T.; Kumar, G.; Li, R.W.; Levi, D.M. Characteristics of fixational eye movements in amblyopia: Limitations on fixation stability and acuity? *Vis. Res.* **2015**, *114*, 87–99. [CrossRef]
19. González, E.G.; Wong, A.M.F.; Niechwiej-Szwedo, E.; Tarita-Nistor, L.; Steinbach, M.J. Eye Position Stability in Amblyopia and in Normal Binocular Vision. *Investig. Opthalmol. Vis. Sci.* **2012**, *53*, 5386–5394. [CrossRef]
20. Koylu, M.T.; Ozge, G.; Kucukevcilioglu, M.; Mutlu, F.M.; Ceylan, O.M.; Akıncıoglu, D.; Ayyıldız, O. Fixation Characteristics of Severe Amblyopia Subtypes: Which One is Worse? *Semin. Ophthalmol.* **2017**, *32*, 553–558. [CrossRef]
21. Wang, S.; Tian, T.; Zou, L.; Wu, S.; Liu, Y.; Wen, W.; Liu, H. Fixation Characteristics of Severe Amblyopia with Eccentric Fixation and Central Fixation Assessed by the MP-1 Microperimeter. *Semin. Ophthalmol.* **2021**, *36*, 360–365. [CrossRef] [PubMed]
22. Scaramuzzi, M.; Murray, J.; Otero-Millan, J.; Nucci, P.; Shaikh, A.G.; Ghasia, F.F. Fixation instability in amblyopia: Oculomotor disease biomarkers predictive of treatment effectiveness. *Prog. Brain Res.* **2019**, *249*, 235–248. [CrossRef] [PubMed]
23. Wang, S.; Zou, L.; Tian, T.; Zhan, A.; Liu, Y.; Wen, W.; Liu, H. Fixation stability improvement after occlusion treatment for severe amblyopia. *Int. Ophthalmol.* **2021**, *42*, 1007–1012. [CrossRef]
24. Carpineto, P.; Ciancaglini, M.; Nubile, M.; Di Marzio, G.; Toto, L.; Di Antonio, L.; Mastropasqua, L. Fixation Patterns Evaluation by Means of MP-1 Microperimeter in Microstrabismic Children Treated for Unilateral Amblyopia. *Eur. J. Ophthalmol.* **2007**, *17*, 885–890. [CrossRef] [PubMed]
25. Maneschg, O.A.; Barboni, M.T.S.; Nagy, Z.Z.; Németh, J. Fixation stability after surgical treatment of strabismus and biofeedback fixation training in amblyopic eyes. *BMC Ophthalmol.* **2021**, *21*, 264. [CrossRef]
26. Balaratnasingam, C.; Inoue, M.; Ahn, S.; McCann, J.; Dhrami-Gavazi, E.; Yannuzzi, L.A.; Freund, K.B. Visual Acuity Is Correlated with the Area of the Foveal Avascular Zone in Diabetic Retinopathy and Retinal Vein Occlusion. *Ophthalmology* **2016**, *123*, 2352–2367. [CrossRef]
27. Gigengack, N.K.; Oertel, F.C.; Motamedi, S.; Bereuter, C.; Duchow, A.; Rust, R.; Bellmann-Strobl, J.; Ruprecht, K.; Schmitz-Hübsch, T.; Paul, F.; et al. Structure–function correlates of vision loss in neuromyelitis optica spectrum disorders. *Sci. Rep.* **2022**, *12*, 17545. [CrossRef]

28. Miki, A.; Hayashida, M.; Inoue, Y.; Yamada, Y.; Nakamura, M. The relationship between foveal avascular zone and fixation stability in patients with a history of retinopathy of prematurity. *Investig. Ophthalmol. Vis. Sci.* **2018**, *59*, 2753.
29. Nourinia, R.; Rajavi, Z.; Sabbaghi, H.; Hassanpour, K.; Ahmadieh, H.; Kheiri, B.; Rajabpour, M. Optical Coherence Tomography Angiography in Patients with Amblyopia. *Strabismus* **2022**, *30*, 132–138. [CrossRef]
30. Ye, H.; Wang, S.; Zhang, Y.; Fang, W.; Ye, H.; Chen, L.; Qiao, T. Microvasculature evaluation of anisometropic amblyopia children by Angio-OCT. *Sci. Rep.* **2023**, *13*, 2780. [CrossRef]
31. Huang, X.; Liao, M.; Li, S.; Liu, L. The effect of treatment on retinal microvasculature in children with unilateral amblyopia. *J. Am. Assoc. Pediatr. Ophthalmol. Strabismus* **2021**, *25*, 287.e1–287.e7. [CrossRef]
32. Yilmaz, I.; Ocak, O.B.; Yilmaz, B.S.; Inal, A.; Gokyigit, B.; Taskapili, M. Comparison of quantitative measurement of foveal avascular zone and macular vessel density in eyes of children with amblyopia and healthy controls: An optical coherence tomography angiography study. *J. Am. Assoc. Pediatr. Ophthalmol. Strabismus* **2017**, *21*, 224–228. [CrossRef]
33. McKee, S.P.; Levi, D.M.; Movshon, J.A. The pattern of visual deficits in amblyopia. *J. Vis.* **2003**, *3*, 380–405. [CrossRef]
34. Crossland, M.D.; Dunbar, H.M.P.B.; Rubin, G.S. Fixation stability measurement using the mp1 microperimeter. *Retina* **2009**, *29*, 651–656. [CrossRef]
35. Shahlaee, A.; Pefkianaki, M.; Hsu, J.; Ho, A.C. Measurement of Foveal Avascular Zone Dimensions and its Reliability in Healthy Eyes Using Optical Coherence Tomography Angiography. *Am. J. Ophthalmol.* **2016**, *161*, 50–55. [CrossRef] [PubMed]
36. Scaramuzzi, M.; Murray, J.; Otero-Millan, J.; Nucci, P.; Shaikh, A.G.; Ghasia, F.F. Part time patching treatment outcomes in children with amblyopia with and without fusion maldevelopment nystagmus: An eye movement study. *PLoS ONE* **2020**, *15*, e0237346. [CrossRef] [PubMed]
37. Molina-Martín, A.; Piñero, D.P.; Pérez-Cambrodí, R.J. Normal Values for Microperimetry with the MAIA Microperimeter: Sensitivity and Fixation Analysis in Healthy Adults and Children. *Eur. J. Ophthalmol.* **2017**, *27*, 607–613. [CrossRef] [PubMed]
38. Morales, M.U.; Saker, S.; Wilde, C.; Pellizzari, C.; Pallikaris, A.; Notaroberto, N.; Rubinstein, M.; Rui, C.; Limoli, P.; Smolek, M.K.; et al. Reference Clinical Database for Fixation Stability Metrics in Normal Subjects Measured with the MAIA Microperimeter. *Transl. Vis. Sci. Technol.* **2016**, *5*, 6. [CrossRef]
39. Mølgaard, I.-L.; Biering-Sørensen, K.; Michelsen, N.; Elmer, J.; Rydberg, A. Amblyopia screening in kindergartens with tno stereotest. *Acta Ophthalmol.* **1984**, *62*, 156–162. [CrossRef]
40. Miladinović, A.; Quaia, C.; Ajčević, M.; Diplotti, L.; Cumming, B.G.; Pensiero, S.; Accardo, A. Ocular-following responses in school-age children. *PLoS ONE* **2022**, *17*, e0277443. [CrossRef]
41. Quaia, C.; FitzGibbon, E.J.; Optican, L.M.; Cumming, B.G. Binocular Summation for Reflexive Eye Movements: A Potential Diagnostic Tool for Stereodeficiencies. *Investig. Opthalmol. Vis. Sci.* **2018**, *59*, 5816–5822. [CrossRef] [PubMed]

life

MDPI

Review

Promising High-Tech Devices in Dry Eye Disease Diagnosis

Andrea De Luca, Alessandro Ferraro, Chiara De Gregorio, Mariateresa Laborante, Marco Coassin, Roberto Sgrulletta and Antonio Di Zazzo *

Ophthalmology Complex Operative Unit, University Campus Bio-Medico, 00128 Rome, Italy
* Correspondence: a.dizazzo@policlinicocampus.it

Abstract: Background: Dry eye disease (DED) is a common and debilitating condition that affects millions of people worldwide. Despite its prevalence, the diagnosis and management of DED can be challenging, as the condition is multifactorial and symptoms can be nonspecific. In recent years, there have been significant advancements in diagnostic technology for DED, including the development of several new devices. Methods: A literature review of articles on the dry eye syndrome and innovative diagnostic devices was carried out to provide an overview of some of the current high-tech diagnostic tools for DED, specifically focusing on the TearLab Osmolarity System, DEvice Hygrometer, IDRA, Tearcheck, Keratograph 5M, Cornea Dome Lens Imaging System, I-PEN Osmolarity System, LipiView II interferometer, LacryDiag Ocular Surface Analyzer, Tearscope-Plus, and Cobra HD Camera. Conclusions: Despite the fact that consistent use of these tools in clinical settings could facilitate diagnosis, no diagnostic device can replace the TFOS algorithm.

Keywords: dry eye disease; diagnostic device; ocular surface

check for updates

Citation: De Luca, A.; Ferraro, A.; De Gregorio, C.; Laborante, M.; Coassin, M.; Sgrulletta, R.; Di Zazzo, A. Promising High-Tech Devices in Dry Eye Disease Diagnosis. *Life* **2023**, *13*, 1425. https://doi.org/10.3390/life13071425

Academic Editors: José-María Sánchez-González, Carlos Rocha-de-Lossada and Alejandro Cerviño

Received: 8 May 2023
Revised: 15 June 2023
Accepted: 19 June 2023
Published: 21 June 2023

1. Introduction

The TFOS DEWS II (Tear Film and Ocular Surface Society International Dry Eye Workshop II, 2017) defines dry eye disease (DED) as "a multifactorial disease of the ocular surface characterized by a loss of homeostasis of the tear film and accompanied by ocular symptoms, in which tear film instability and hyperosmolarity, ocular surface inflammation and damage, and neurosensory abnormalities play etiological roles" [1]. The prevalence of DED exhibits a positive correlation with advancing age and varies between five percent and fifty percent across the overall population [2]. DED is characterized by a range of symptoms such as ocular pain, burning, stinging, discomfort, a foreign body sensation, poor visual acuity, photophobia, and irritation [1,2]. The symptoms of dry eye disease can range from minor discomfort to substantial grievances that interfere with daily functioning, decrease quality of life, and carry notable consequences for the socioeconomic structure [1,2].

The first phase in the process of diagnosing dry eye disease involves the utilization of triage questions, which could establish the need for additional DED evaluation and exclude disorders such as conjunctivitis, blepharitis, Sjögren syndrome, infection, and lid-related disease.

Based on the TFOS DEWS II, a dry eye diagnosis requires the patient to score positively on one of two specific symptom questionnaires (DEQ-5, Dry Eye Questionnaire-5 score ≥ 6 or OSDI, Ocular Surface Disease Index score ≥ 13). This must be followed by the presence of a minimum of one positive clinical sign, such as decreased tear film stability (NIBUT, non-invasive tear break-up time, <10 s), elevated tear osmolarity (>308 mOsm/L), significant inter-eye disparity in osmolarity (>8 mOsm/L), or ocular surface damage indicated by fluorescein and lissamine green (>5 corneal spots, >9 conjunctival spots, or lid margin ≥ 2 mm length and $\geq$25% width).

OSDI refers to a verified questionnaire that is commonly employed in clinical trials due to its ability to provide a rapid assessment of dry eye disease (DED) and its impact on the quality of life (QoL) of patients [3]. The OSDI consists of 12 items that aim to evaluate the symptoms experienced by patients during the preceding week. It is organized into three sections: the occurrence of symptoms, the impact on vision-related quality of life, and the identification of any environmental triggers [3]. Each item is evaluated using a scale ranging from zero to four. The scale used to measure frequency is as follows: none of the time (0), some of the time (1), half of the time (2), the majority of the time (3), and the entire time (4). The overall numerical value is calculated utilizing a range that extends from zero to one hundred, whereby elevated scores denote heightened levels of impairment. A score of 0 to 12 is regarded as normal; a score of 13 to 22 suggests mild disease; a score of 23 to 32 denotes moderate DED; and a score of 33 to 100 indicates severe DED [3]. Furthermore, OSDI does not discriminate between evaporative and aqueous deficiencies [4].

It is possible that the TFOS DEWS II diagnostic algorithm is not the most effective method when used in a clinical environment, despite the fact that it provides both a comprehensive and full approach to detecting dry eye disease. Although it offers a complete evaluation, the diagnostic procedure typically consists of a number of steps that can be challenging to carry out in fast-paced clinical settings with limited time.

Therefore, in this paper, we examine new high-tech imaging systems for ocular surface evaluation. These systems claim to have several benefits over traditional methods of diagnosis, such as being non-invasive, providing standardized and objective results, enabling the monitoring of disease progression and treatment effectiveness, being user-friendly, and enabling rapid task execution. However, regardless of the studies reviewed, the reliability of these devices is low.

2. Materials and Methods

A literature review of articles on dry eye syndrome and innovative diagnostic devices published in the last 15 years and available on the National Library of Medicine was carried out without any restriction of language, especially focusing on the TearLab Osmolarity System, IDRA, Tearcheck, Keratograph 5M, I-Pen Osmolarity System, Lipiview II Interferometer, LacryDiag Ocular Surface Analyzer, Cornea Dome Lens Imaging System, DEvice Hygrometer, Tearscope-Plus, and Cobra HD Camera. All published peer-reviewed randomized clinical trials, meta-analyses, systematic reviews, and observational studies about these topics were evaluated. Table 1 shows all the devices analyzed and the exams they perform.

TearLab Osmolarity System® (TearLab Corporation, San Diego, CA, USA) is a non-invasive diagnostic device that analyzes the osmolarity of a patient's tears. Osmolarity refers to the total concentration of dissolved substances present in a solution without regard to their density, size, molecular weight, or electrical charges. The process of tears evaporating, a decrease in the production of tears, and the dysfunction of the meibomian gland are all factors that contribute to an elevation in tear osmolarity. The evaluation of tear osmolarity is considered a highly effective diagnostic tool for every type of dry eye syndrome [5]. The tear film of the exposed ocular surface area exhibits a lower osmolarity. Various environmental factors such as wind, cigarette smoke, indoor air conditioning, and heat, as well as prolonged computer use leading to reduced blinking frequency, have been identified as potential impediments to evaporation, thereby affecting tear osmolarity [6]. Tear osmolarity was also found to correlate with increased concentrations of inflammatory cytokines and matrix metalloproteinases (MMPs), as well as HLA-DR (Human Leukocyte Antigen-DR) overexpression, suggesting that tear osmolarity could potentially serve as a predictive measure for ocular surface diseases that are linked to high levels of inflammatory mediators [6].

Table 1. Ocular surface diagnostics devices comparison.

Device	Main Exams Performed											
	NIBUT	Osmolarity	LLT	TMH	Meibography	Eye Blink	Bulbar Redness	Inflammatory Evaluation	Demodex Presence	OSDI	Accurate Ocular Surface Images	RH
TearLab Osmolarity System®		x										
IDRA®	x		x	x	x	x						
Tearcheck®	x		x	x	x	x	x	x	x	x		
Keratograph 5M®	x		x	x	x		x					
I-PEN Osmolarity System®		x										
LipiView®II interferometer			x		x	x						
LacryDiag Ocular Surface Analyzer	x		x	x	x							
Cornea Dome Lens Imaging System®											x	
DEvice Hygrometer©												x
Tearscope-Plus™	x		x	x								
Cobra HD Camera					x							

NIBUT, non-invasive tear break-up time; LLT, lipid layer thickness; TMH, tear meniscus height; RH, relative humidity.

The TearLab Osmolarity System is composed of a set of instruments, including a reader device, pens, and test cards. The reader is a small countertop device that calculates and shows the osmolarity test results on a liquid crystal display. The TearLab osmolarity device is equipped with a pair of pens that are used to hold the test card and transmit the data to the reader. The test card is attached to the pens and takes a sample of 50 nL in milliosmoles per liter (mOsm/L) units. The contact between the tear film and the eyelid occurs at the temporal margin. After hearing the beep confirming successful tear collection, the pen is docked into the reader [7].

This osmometer has several advantages, including being a small, portable device that can be used in a doctor's office and requiring less than 100 nL of tear fluid [5]. Furthermore, it uses a temperature-corrected tear fluid impedance measurement, enabling an indirect evaluation of tear osmolarity and providing precise results within a brief timeframe [5,8]. It may also be used in combination with other diagnostic tools, such as the LipiView system, to provide a more detailed image of a patient's dry eye condition. According to Szczesna-Iskander, it is necessary to take a minimum of three consecutive measurements to obtain clinically trustworthy tear osmolarity values using the TearLab Osmolarity System. The utilization of the highest osmolarity value to identify DED requires careful consideration due to the frequent occurrence of anomalous readings of tear osmolarity [9]. Nevertheless, Szalai et al. found significant overlap in tear osmolarity values measured with the TearLab system in the control (22 healthy individuals) and dry eye groups (21 patients with non-Sjögren syndrome dry eye (NSSDE) and 20 patients with Sjögren syndrome dry eye (SSDE)), implying that measurement of tear osmolarity utilizing the TearLab osmometer is highly variable and does not distinguish individuals diagnosed with dry eye disease from healthy controls (mean tear osmolarity was 296.77 $\pm$ 16.48 mOsm/L in NSSDE, 303.36 $\pm$ 17.22 mOsm/L in SSDE, and 303.52 $\pm$ 12.92 mOsm/L in the control group; $p = 0.018$) [10].

As a result, TearLab is a quick tool in clinical practice, but its usefulness is limited by the literature-reported low reliability in recognizing DED and its primary use in evaporative dry eye.

IDRA® Ocular Surface Analyzer (SBM SISTEMI, Inc., Torino, Italy) is a diagnostic tool that uses infrared and ultraviolet light to evaluate the health of the ocular surface. Changes in the tear film and meibomian glands, which can be indicative of DED, can be detected by the instrument. The instrument is able to identify MG as well as all three layers of the tear film (lipid, aqueous, and mucin). This allows physicians to determine which parts of the tear film require treatment based on the type of insufficiency. IDRA must be placed between a slit lamp and a biomicroscope. Its pins have been designed to fit perfectly into the hole left by removing the plate used for the tonometer, and it conducts a 5-minute non-invasive test [11]. The instrument produces a beam of white light onto the corneal surface, and the resultant reflection of light from the tear film presents a white, fan-shaped region that covers the inferior third of the cornea [12]. The five parameters include NIBUT, TMH, lipid layer interferometry, ocular blink quality, and infrared meibography. The non-invasive break-up time (NIBUT) is determined through the utilization of Placido's disc to project ring patterns on the cornea, followed by the measurement of the duration in seconds between the complete blink and the first disturbance of the reflected image on the cornea [11]. The utilization of infrared illumination in non-invasive meibography has the capability to identify morphological alterations in the meibomian gland. On the other hand, tear interferometry can be employed to assess the lipid layer of the tear film. The evaluation of meibomian gland morphology offers valuable clinical evidence for the diagnosis of evaporative DED, while assessments of the lipid layer of the tear film facilitate the monitoring of meibomian gland function [10]. The photograph shows the identification of the upper and lower tear meniscus as well as the evaluation of tear meniscus height along the lower lid margin [11]. A prospective study with 75 patients (40 with DED and 35 healthy subjects) found good concordance between the IDRA ocular surface analyzer and standard diagnostic procedures in differentiating between individuals with normal ocular function and those with dry eye disease. It had an area under the curve of 0.868 (95% confidence interval: 0.809–0.927) to detect DED [13].

The lipid layer is important for regulating the evaporation of the tear film. A test for lipid layer pattern (LLP) evaluation is based on interference phenomena, but it is influenced by subjective interpretation of the patterns [14]. The lipid layer thicknesses are determined through the utilization of Dr. Guillon's international grading system, which enables the calculation of the average, maximum, and minimum thicknesses of the lipid layer pattern grades [15]. The grades are converted to nanometer units and classified into a range of 15 to 100 nm according to the observed patterns. The maximum cutoff wavelength of IDRA is 100 nm [16].

Despite the fact that IDRA has the benefit of assessing multiple ocular surface parameters with a single device, the findings are contradictory. In a retrospective cross-sectional study with 47 non-Sjögren dry eye patients, Jeon et al. demonstrated a significant correlation between dry eye symptoms and the partial blink rate as well as meibomian gland dropout rates measured using IDRA. On the other hand, Lee et al. demonstrated that IDRA exhibited a considerably lower percentage of meibomian gland dropout and a greater partial blink rate than another device, LipiView® II, in a cross-sectional, single-visit observational study with 47 participants [11,12]. These findings suggest that these devices should not be used interchangeably when assessing meibomian gland dropouts and partial blink rates [11]. Rinert et al. demonstrated a favorable correlation between everyday clinical diagnostic examinations. The researchers found that the utilization of pathologic meibography, interferometry, and tear meniscus measurements with the analyzer produced a 96% estimated probability of dry eye disease. Simultaneously, the percentage of eyes exhibiting pathological observations in the three sets of examinations was relatively minimal, suggesting that dry eye disease (DED) may manifest in diverse clinical presentations and necessitate a comprehensive assessment [13].

Thus, IDRA appears to be an interesting device for diagnosing DED because it evaluates all three components of the tear film; however, the literature presents conflicting and limited findings.

Tearcheck® (E-Swin, Houdan, France) is a stand-alone device with an integrated screen that allows the user to view all acquisitions and exams in real time. The device facilitates quick evaluations that include nine examinations: non-invasive break time, tear film stability, ocular surface inflammatory assessment, meibography IR, Demodex, eye redness, abortive blinking, tear meniscus height, and the OSDI questionnaire. This results in a simple dry eye analysis.

Using the Demodex exam, an enlarged image of the base of the eyelashes can be obtained, allowing for the tracing and visualization of signs indicating the presence of Demodex mites. As a result, the device takes high-resolution images of the ocular surface, enabling the detection of changes in the cornea, conjunctiva, and tear film, such as the existence of inflammation or ocular surface damage (dry spots or erosion).

Although the device is non-invasive and simple to use, there is no evidence in the literature to support its use in the diagnosis and management of DED.

Keratograph 5M® (Oculus, Wetzlar, Germany) is a diagnostic device that uses non-invasive imaging technology to evaluate the health of the cornea. The device can identify changes in the surface and shape of the cornea, which can be indicative of DED. It is a corneal topographer that is equipped with a real keratometer and a color camera. Its purpose is to capture external images by projecting a ring pattern from a placido disc on the tear film surface using an infrared light source. It may evaluate non-invasive break-up time (NIBUT), meibography, bulbar redness, tear meniscus height, lipid layer (interferometry), and tear film dynamics (monitoring of tear film particle flow, from which inferences about tear film viscosity can be inferred).

The findings of this tool in the literature are also contradictory. On the one hand, the keratograph has significant examiner bias [17], but on the other hand, it has been reported to have strong inter-examiner reproducibility (mean difference between examiners of 0.08 ± 0.55 and 0.13 ± 0.50 grade units in two separate sessions, respectively) with low within-subject variability (95% limits of agreement for two different examiners of −1.02 to +1.10 and −1.27 to +1.09 grade units, respectively) [18]. In a prospective study with 42 patients with DED and 42 healthy subjects, Tian et al. found that utilizing the non-invasive Keratograph tear break-up time (NIKBUT) and tear meniscus height (TMH) measurements through the K5M device could serve as a straightforward and non-invasive method for screening dry eye while also exhibiting satisfactory levels of repeatability and reproducibility (coefficient of variation (CV%) $\leq$ 26.1% and intraclass correlation coefficient (ICC) $\geq$ 0.75 for all measurements). Nevertheless, it was observed that NIKBUTs exhibited greater reliability in individuals with dry eye disease (DED) as compared to TMH [19].

Sutphin et al. concluded that keratographic measures cannot be considered interchangeable alternatives for commonly used clinical measures. Furthermore, they found that there is no specific test that can provide objective support for the diagnosis [20]. Indeed, according to Pérez-Bartolomé et al., the Keratograph 5M was observed to overestimate ocular redness scores in comparison to subjective grading scales when utilized for the purpose of assessing the degree of ocular redness [21]. Furthermore, in an observational cross-sectional study of 47 subjects with DED and 41 normal control subjects, Chen et al. showed a limited association between the keratograph tear meniscus height (TMH) and Schirmer scores among individuals with dry eye disease (DED). They also demonstrated that in the comparison of Fourier-domain optical coherence tomography (FD-OCT) and the Keratograph 5M, both instruments exhibited notable diagnostic precision in distinguishing between normal patients and those with dry eye disease. However, it was observed that the FD-OCT measurements of tear meniscus height (TMH) were more dependable than the keratograph data in the DED group [22]. Specifically, while the keratograph and FD-OCT measurements of TMH were closely correlated, the former tended to yield lower measurements than the latter [23].

Despite the fact that the Keratograph 5M is a non-invasive diagnostic topographer for DED, it is not yet a substitute for multiple clinical tests such as the Schirmer test and FBUT because its reliability is very weak.

I-PEN Osmolarity System® (I-MED Pharma Inc., Dollard-des-Ormeaux, QC, Canada) is a portable electrical tool that assesses the osmolarity of tears by measuring the electrical impedance of the eye tissues on the palpebral conjunctival membrane. The occurrence of an inflammatory cascade at the ocular surface is initiated by tear film hyperosmolarity, which ultimately leads to the loss of goblet cells, epithelial injury, and the production of cytokines. This condition is responsible for causing ocular distress and vision impairment in patients with DED. The I-PEN's usefulness has been questioned in the literature, though numerous studies have considered the I-Pen appropriate and reliable for clinical use [24–26]. Park et al. concluded that the I-Pen osmometer demonstrates favorable performance in diagnosing DED in clinical settings; however, it should not be solely relied upon for evaluating DED. Nonetheless, the I-Pen osmometer can serve as a valuable tool for screening and identifying dry eye disease [24]. In contrast, some researchers failed to establish any correlations between tear film osmolarity values acquired through the I-PEN system and various subjective or objective parameters of dry eye disease (DED). Furthermore, they showed that the I-PEN system was less effective than the TearLab Osmolarity System in delineating subjects with and without dry eye disease [7,27,28]. Shimazaki et al. found no statistically significant difference in mean tear film osmolarity between the DED (871 eyes) and non-DED (51 eyes) groups using the I-PEN system (294.76 ± 16.39 vs. 297.76 ± 16.72 mOsms/L, respectively, p = 0.32). Furthermore, motion may affect osmolarity readings acquired through the I-Pen system; however, the influence of this factor can be minimized if the measurements are carried out by a highly trained clinician [29]. Alanazi et al. evaluated the relationship between osmolarity results acquired by the TearLabTM and I-Pen® systems in individuals with a high body mass index (BMI). The I-Pen® results (294–336 mOsm/L in the study group of 30 male subjects with a high BMI and 278–317 mOsm/L in the control group of 30 healthy males) were significantly higher than the TearLab™ scores (278–309 mOsm/L in the study group and 263–304 mOsm/L in the control group). Furthermore, the outcomes obtained from the I-Pen® measurements exhibited significant variations in osmolarity values and demonstrated a considerable lack of accuracy in distinguishing normal eyes as compared to the TearLab™ system [29].

As a result, the I-PEN is a portable, easy-to-use, autocalibrated device that is not affected by variations in tear film volume and requires only a simple touch of the palpebral conjunctiva. Nevertheless, it can only detect tear osmolarity, and further investigations are required to determine its utility.

LipiView® II interferometer (TearScience Inc., Morrisville, NC, USA) is a device for ocular imaging that examines the interferometric pattern of the tear film. It achieves this by measuring the lipid layer thickness (LLT) of the tear film with nanometer accuracy; however, it has an upper cut-off to assess LLT values of 100 nm [11]. In addition to that, it records the dynamics of blinking and images the structure of the meibomian gland. Compared to IDRA, in a cross-sectional single-visit observational study with 47 participants, Min Lee et al. found no significant difference in LLT. However, IDRA had a considerably lower rate of MG dropout and a higher PBR (IDRA MG dropout 45.36 ± 21.87 and PBR 0.23 ± 0.27; LipiView® II interferometer MG dropout 36.51 ± 17.53 and PBR 0.51 ± 0.37) [11]. In comparison to Keratograph 5M (K5M), Wong et al. showed that LVII exhibited a statistically significant reduction in meiboscores and a lower percentage of MG dropout in 20 subjects (1.43 ± 0.78 vs. 1.90 ± 0.81, p = 0.001) [30]. These results suggest that there is poor interchangeability between the methods used to evaluate DED features, particularly MG dropouts and PBR.

In conclusion, the data obtained from the LVII LLT should be compared to other instruments. However, additional studies with larger sample sizes are necessary.

LacryDiag Ocular Surface Analyzer (Quantel Medical, Cournon-d'Auvergne, France) is an ophthalmic device used to diagnose and monitor the tear film and meibomian glands. It takes non-invasive photographs with white or infrared light to assess the height of the lower tear meniscus, the distance between the upper and lower eyelids, tear film interferometry, and non-invasive tear film break-up time. Toth et al. demonstrated that it is a non-invasive, simple-to-use device able to analyze the tear film and save photos for later use [31]. Despite this result, there is a great deal of variability between measurements performed by this instrument and those performed by another innovative device, such as the OCULUS Keratograph 5M, possibly reflecting differences in image processing or the need for subjective evaluation by the observer for a considerable number of these measurements [32,33]. Ward et al., for instance, evaluated the repeatability of the LacryDiag Ocular Surface Analyzer for both intra- and inter-observer measurements and compared it to the OCULUS Keratograph 5M in 30 healthy subjects. Their findings revealed a good relationship between the devices (no differences in mean values for tear meniscus height, NIBUT, or tear film interferometry, except for lipid layer interferometry), but low agreement for any individual observer (intra-observer variability for NIBUT was significantly higher for the Keratograph, p = 0.0003 for observer A and p < 0.0001 for observer B). According to the authors, the observed inconsistency could be attributed to the utilization of repeated testing and the inclusion of subjects without dry eye conditions. Therefore, the authors concluded that in the identification and follow-up of patients with dry eye disease (DED), it is essential to consider the reproducibility of the testing instrument and the utilization of different outcome measures [33].

In summary, LacryDiag is a promising instrument for assessing the ocular surface, but there is a lack of research about it in the medical literature.

Cornea Dome Lens Imaging System® (Occyo, Innsbruck, Austria) is an imaging system that attempts to provide uniform ocular surface color photographs respecting position, illumination, focus, and operator independence. This is achieved by overcoming the limitations of objective methods that are based on digital ocular surface images. The device is composed of a novel imaging lens that conforms accurately to the curvature of the eye, enabling high-resolution imaging of the visible ocular surface. In addition, the device incorporates a fixation target that guarantees a centralized view into the lens, thereby minimizing eye movements. Furthermore, the set is composed of software designed for eye tracking as well as an illumination unit. To maintain the stability of the patient's head, a chin rest and a forehead band are utilized. The aim of the system is to obtain photographs in a standardized manner without the need for human intervention. Therefore, according to Lins, this tool possesses the ability to evaluate the extent of bulbar eye redness in an objective and replicable manner, utilizing the images it has captured [34].

This technology has the potential to provide an objective technique based on the digital ocular surface for assessing bulbar eye redness, overcoming the limitations of subjective photographic scales that suffer from inter-rater variability. However, its role in the DED diagnostic process should be investigated in future studies.

DEvice Hygrometer© (AI, Rome, Italy) is a complete, low-cost diagnostic-therapeutic tool for ocular surface management that has the capability to quickly identify the entities of production, clearance, and stability, along with the severity of tear film evaporation, and drives the subsequent therapy through the utilization of simple algorithms. The device operates by detecting changes in the relative humidity (RH) levels within a confined environment surrounding the ocular surface (Figure 1).

The diagnostic component of the device works by measuring the evaporation of the tear film from the ocular surface at a variable speed that may be modified by the user. At a certain temperature, the device measures the baseline and post-stimulus relative humidity values. The sensor is placed in a cup on the orbital edges by the operator. The measurements are carried out in a closed environment made up of the cup and the ocular surface system. Using the acquired data, it is possible to construct progression curves for relative humidity (RH) that have been corrected for temperature. The collected data are "basal" values that are combined with measurements taken in reaction to diverse kinds of stimuli, such as air blows, alterations in temperature within the microenvironment surrounding the ocular surface, light stimuli, and so on. Additionally, a non-contact sample mechanism built into the device allows for the collection of a certain amount of tear evaporation. Despite the fact that incomplete blinking and tear clearance may have an impact on the accuracy of measurements obtained from the DEvice©, it represents a low-cost, efficient, accurate, rapid, and safer (as it is non-contact) instrument for measurement of tear films. Indeed, a preliminary observational pilot study with 8 patients (2 with DED and 6 healthy subjects) has shown that individuals with dry eye disease (DED) showed higher relative humidity values compared to healthy individuals. However, additional studies involving a larger sample size are necessary to confirm these findings. The diagnostic device exhibits potential for local drug nebulization, thereby presenting an option for alternative therapeutic applications in the future [35].

Thus, even though it is a promising diagnostic tool, its use in clinical practice is now limited to evaporative dry eye.

Tearscope-Plus™ (Keeler, Windsor, UK) is a relatively new portable device that may be connected to a slit lamp for conducting non-invasive assessments of the tear film. It allows the examination of the interference patterns of the lipid layer across the whole cornea without the need for fluorescein, thereby enabling the evaluation of non-invasive break-up time (NIBUT), tear meniscal height (TMH), and lipid layer thickness of the tear film (LLT). Guillon's classification is employed for the purpose of determining the thickness of the lipid layer (LLP, lipid layer pattern). The system includes five different types of lipid layer patterns: open meshwork (OM), closed meshwork (CM), wave (W), amorphous (AM), and color fringe (CO). In addition to normal LLPs and events, atypical ones were described. This method is effective for investigating the quality and structure of the tear film; nevertheless, it is dependent on the observer's judgment, which can be affected by the sort of pattern that is viewed [36]. Visualizing thicker lipid layers can be challenging due to the lack of distinguishing morphological traits and color fringes. Additionally, the subjective perception of the observer can influence the findings. García-Resúa et al. showed that, although there was a significant correlation between classifications made by experienced observers based on Guillon's schema, misinterpretations of the patterns might still occur, even within the same observer [14].

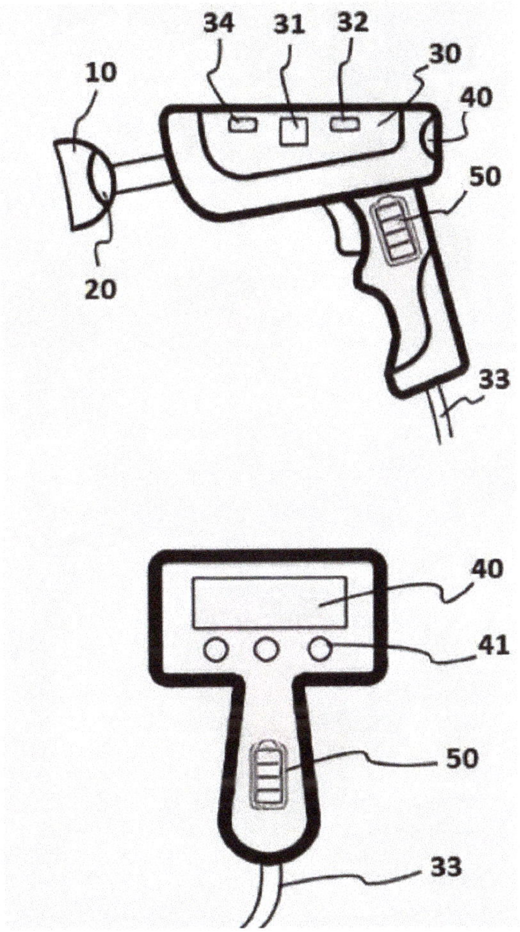

Figure 1. A schematic design of the diagnostic tool system. The figure shows a side and rear view of the schematic representation of a monosensor diagnostic prototype with: an eyepiece cup (10); the sensor (20) placed inside it; a processing board (30) equipped with processor (31), memory (34), and wireless connection device (32); the optional connection cable (33); a digital screen (40) with buttons (41); and a rechargeable power supply battery (50) placed in the handle [35].

Despite the fact that Fodor et al. demonstrated that lower tear meniscus height measurements were more repeatable with Tearscope than slit-lamp biomicroscopy without staining in 31 healthy individuals, the subjectivity inherent in its use limits both its repeatability and its utility in comparison to other instruments that are more objective (Oculus® Keratograph 5M and LipiView®) [37,38].

Finally, Tearscope presents a reliable and consistent method of increasing clinical observation and identification of ocular physiological alterations; yet, this automated procedure is susceptible to human error.

Cobra HD Camera (CSO, Florence, Italy) is a non-mydriatic digital fundus device with modules designed for retinal screening analysis. Additionally, it includes a dedicated module for meibography [17].

In a study conducted by Pult, the association between age, sex, and dry eye symptoms, as well as the quantification of meibomian gland (MG) loss, was investigated using a Cobra fundus camera. The study involved 112 participants and revealed substantial standard deviations in the mean MG loss between participants with and without dry eye symptoms (30 ± 17% and 45 ± 18%, respectively) [39].

Iphra and Gantz investigated the inter-session repeatability (ISR), inter-examiner reproducibility (IER), and within-subject variability (WSV) of the Cobra HD fundus camera meibographer. This study utilized Phoenix software. Participants were classified as either symptomatic or asymptomatic for dry eye based on their Ocular Surface Disease Index (OSDI) questionnaire scores. To determine the IER, seventy-four participants were evaluated on the same day by two examiners, referred to as Examiner 1 and Examiner 2. Subsequently, sixty-six of these participants were re-examined by Examiner 1 on a different date to calculate the ISR. The results showed that the Cobra HD fundus camera meibographer had good repeatability and reproducibility, and clinically similar findings should be obtained when used by different examiners on different occasions [40].

In conclusion, although the Cobra HD camera can only detect meibomian gland loss, it is useful for the meibographic assessment and follow-up of DED progression.

3. Discussion

Easy and rapid dry eye disease diagnosis is still a challenging unmet need in opthalmology. The algorithms proposed by various international societies and committees are frequently time-consuming and costly, and their use in the context of a busy medical setting is limited. Although 20% of our patients suffer from DED and ocular discomfort impact almost 40% of surgeons practice, a proper DED diagnostic method is still missed [41–43]. Therefore, several new diagnostic tools aim to fill this gap, making the diagnosis with a single "click", although at a higher cost.

The consistent use of these instruments in clinical settings may facilitate the diagnosis, tracking, and prevention of ocular surface diseases such as dry eye syndrome. However, the results in the literature are few and inconsistent. Most likely, there is no clear way for practitioners to use these automated tools to diagnose dry eyes in a standardized way. Furthermore, the lack of intra- and interobserver repeatability in certain measurement instruments limits neutrality and increases bias, influencing their use and distribution. A potential drawback of these imaging systems is their high cost, which can limit their accessibility in many healthcare centers.

4. Conclusions

No diagnostic instrument can replace the complex TFOS algorithm, and there is no single tool for a specific diagnosis, but research in this field is very active, and such primordial devices may be a promising reality in the very near future.

Funding: This research received no external funding.

Institutional Review Board Statement: Not applicable.

Informed Consent Statement: Not applicable.

Data Availability Statement: Data sharing not applicable.

Conflicts of Interest: The authors declare no conflict of interest.

References

1. Craig, J.P.; Nichols, K.K.; Akpek, E.K.; Caffery, B.; Dua, H.S.; Joo, C.-K.; Liu, Z.; Nelson, J.D.; Nichols, J.J.; Tsubota, K.; et al. TFOS DEWS II Definition and Classification Report. *Ocul. Surf.* **2017**, *15*, 276–283. [CrossRef] [PubMed]
2. Stapleton, F.; Alves, M.; Bunya, V.Y.; Jalbert, I.; Lekhanont, K.; Malet, F.; Na, K.-S.; Schaumberg, D.; Uchino, M.; Vehof, J.; et al. TFOS DEWS II Epidemiology Report. *Ocul. Surf.* **2017**, *15*, 334–365. [CrossRef] [PubMed]
3. Hashmani, N.; Munaf, U.; Saleem, A.; Javed, S.O.; Hashmani, S. Comparing SPEED and OSDI Questionnaires in a Non-Clinical Sample. *Clin. Ophthalmol.* **2021**, *15*, 4169–4173. [CrossRef] [PubMed]

4. Machalińska, A.; Zakrzewska, A.; Safranow, K.; Wiszniewska, B.; Machaliński, B. Risk Factors and Symptoms of Meibomian Gland Loss in a Healthy Population. *J. Ophthalmol.* **2016**, *2016*, 7526120. [CrossRef]
5. Srinivasan, S.; Nichols, K.K. Editorial: Collecting tear osmolarity measurements in the diagnosis of dry eye. *Expert Rev. Ophthalmol.* **2009**, *4*, 451–453. [CrossRef]
6. Versura, P.; Campos, E.C. TearLab® Osmolarity System for diagnosing dry eye. *Expert Rev. Mol. Diagn.* **2013**, *13*, 119–129. [CrossRef]
7. Tavakoli, A.; Markoulli, M.; Flanagan, J.; Papas, E. The validity of point of care tear film osmometers in the diagnosis of dry eye. *Ophthalmic Physiol. Opt.* **2022**, *42*, 140–148. [CrossRef]
8. Benelli, U.; Nardi, M.; Posarelli, C.; Albert, T.G. Tear osmolarity measurement using the TearLab Osmolarity System in the assessment of dry eye treatment effectiveness. *Contact Lens Anterior Eye* **2010**, *33*, 61–67. [CrossRef]
9. Szczesna-Iskander, D.H. Measurement variability of the TearLab Osmolarity System. *Contact Lens Anterior Eye* **2016**, *39*, 353–358. [CrossRef]
10. Szalai, E.; Berta, A.; Szekanecz, Z.; Szûcs, G.; Módis, L. Evaluation of tear osmolarity in non-Sjögren and Sjögren syndrome dry eye patients with the TearLab system. *Cornea* **2012**, *31*, 867–871. [CrossRef]
11. Lee, J.M.; Jeon, Y.J.; Kim, K.Y.; Hwang, K.-Y.; Kwon, Y.-A.; Koh, K. Ocular surface analysis: A comparison between the LipiView® II and IDRA®. *Eur. J. Ophthalmol.* **2021**, *31*, 2300–2306. [CrossRef]
12. Jeon, Y.J.; Song, M.Y.; Kim, K.Y.; Hwang, K.-Y.; Kwon, Y.-A.; Koh, K. Relationship between the partial blink rate and ocular surface parameters. *Int. Ophthalmol.* **2021**, *41*, 2601–2608. [CrossRef]
13. Rinert, J.M.; Branger, G.; Bachmann, L.M.M.; Pfaeffli, O.; Iselin, K.; Kaufmann, C.; Thiel, M.A.M.; Baenninger, P.B. Accuracy of a New Noninvasive Automatic Ocular Surface Analyzer for the Diagnosis of Dry Eye Disease-Two-Gate Design Using Healthy Controls. *Cornea* **2023**, *42*, 416–422. [CrossRef]
14. García-Resúa, C.; Pena-Verdeal, H.; Miñones, M.; Giráldez, M.J.; Yebra-Pimentel, E. Interobserver and intraobserver repeatability of lipid layer pattern evaluation by two experienced observers. *Contact Lens Anterior Eye* **2014**, *37*, 431–437. [CrossRef]
15. Guillon, J.P. Non-invasive tearscope plus routine for contact lens fitting. *Contact Lens Anterior Eye* **1998**, *21* (Suppl. S1), S31–S40. [CrossRef]
16. Zhao, Y.; Tan, C.L.S.; Tong, L. Intra-observer and inter-observer repeatability of ocular surface interferometer in measuring lipid layer thickness. *BMC Ophthalmol.* **2015**, *15*, 53. [CrossRef]
17. Garduño, F.; Salinas, A.; Contreras, K.; Rios, Y.; García, N.; Quintanilla, P.; Mendoza, C.; Leon, M.G. Comparative study of two infrared meibographers in evaporative dry eye versus nondry eye patients. *Eye Contact Lens* **2021**, *47*, 335–340. [CrossRef]
18. Ngo, W.; Srinivasan, S.; Schulze, M.; Jones, L. Repeatability of grading meibomian gland dropout using two infrared systems. *Optom. Vis. Sci.* **2014**, *91*, 658–667. [CrossRef]
19. Tian, L.; Qu, J.-H.; Zhang, X.-Y.; Sun, X.-G. Repeatability and Reproducibility of Noninvasive Keratograph 5M Measurements in Patients with Dry Eye Disease. *J. Ophthalmol.* **2016**, *2016*, 8013621. [CrossRef]
20. Sutphin, J.E.; Ying, G.-S.; Bunya, V.Y.; Yu, Y.; Lin, M.C.; McWilliams, K.; Schmucker, E.; Kuklinski, E.J.; Asbell, P.A.; Maguire, M.G.; et al. Correlation of Measures from the OCULUS Keratograph and Clinical Assessments of Dry Eye Disease in the Dry Eye Assessment and Management Study. *Cornea* **2022**, *41*, 845–851. [CrossRef]
21. Pérez-Bartolomé, F.; Sanz-Pozo, C.; Martinez-De-La-Casa, J.M.; Arriola-Villalobos, P.; Fernández-Pérez, C.; García-Feijoó, J. Assessment of ocular redness measurements obtained with keratograph 5M and correlation with subjective grading scales. *J. Fr. Ophtalmol.* **2018**, *41*, 836–846. [CrossRef] [PubMed]
22. Chen, M.; Wei, A.; Xu, J.; Zhou, X.; Hong, J. Application of Keratograph and Fourier-Domain Optical Coherence Tomography in Measurements of Tear Meniscus Height. *J. Clin. Med.* **2022**, *11*, 1343. [CrossRef]
23. Baek, J.; Doh, S.H.; Chung, S.K. Comparison of Tear Meniscus Height Measurements Obtained with the Keratograph and Fourier Domain Optical Coherence Tomography in Dry Eye. *Cornea* **2015**, *34*, 1209–1213. [CrossRef] [PubMed]
24. Park, J.; Choi, Y.; Han, G.; Shin, E.; Han, J.; Chung, T.-Y.; Lim, D.H. Evaluation of tear osmolarity measured by I-Pen osmolarity system in patients with dry eye. *Sci. Rep.* **2021**, *11*, 7726. [CrossRef] [PubMed]
25. Fagehi, R.; Al-Bishry, A.; Alanazi, M.; Abusharha, A.; El-Hiti, G.; Masmali, A. Investigation of the repeatability of tear osmolarity using an I-PEN osmolarity device. *Taiwan J. Ophthalmol.* **2020**, *11*, 168–174. [CrossRef] [PubMed]
26. Chan, C.C.; Borovik, A.; Hofmann, I.; Gulliver, E.; Rocha, G. Validity and Reliability of a Novel Handheld Osmolarity System for Measurement of a National Institute of Standards Traceable Solution. *Cornea* **2018**, *37*, 1169–1174. [CrossRef]
27. Shimazaki, J.; Sakata, M.; Den, S.; Iwasaki, M.; Toda, I. Tear Film Osmolarity Measurement in Japanese Dry Eye Patients Using a Handheld Osmolarity System. *Diagnostics* **2020**, *10*, 789. [CrossRef]
28. Nolfi, J.; Caffery, B. Randomized comparison of in vivo performance of two point-of-care tear film osmometers. *Clin. Ophthalmol.* **2017**, *11*, 945–950. [CrossRef]
29. Alanazi, M.; El-Hiti, G.; Alhafy, N.; Almutleb, E.; Fagehi, R.; Alanazi, S.; Masmali, A. Correlation between osmolarity measurements using the TearLab™ and I-Pen® systems in subjects with a high body mass index. *Adv. Clin. Exp. Med.* **2022**, *31*, 1413–1418. [CrossRef]
30. Wong, S.; Srinivasan, S.; Murphy, P.J.; Jones, L. Comparison of meibomian gland dropout using two infrared imaging devices. *Contact Lens Anterior Eye* **2019**, *42*, 311–317. [CrossRef]

31. Tóth, N.; Szalai, E.; Rák, T.; Lillik, V.; Nagy, A.; Csutak, A. Reliability and clinical applicability of a novel tear film imaging tool. *Graefes Arch. Clin. Exp. Ophthalmol.* **2021**, *259*, 1935–1943. [CrossRef]
32. Garcia-Terraza, A.L.; Jimenez-Collado, D.; Sanchez-Sanoja, F.; Arteaga-Rivera, J.Y.; Flores, N.M.; Pérez-Solórzano, S.; Garfias, Y.; Graue-Hernández, E.O.; Navas, A. Reliability, repeatability, and accordance between three different corneal diagnostic imaging devices for evaluating the ocular surface. *Front. Med.* **2022**, *9*, 893688. [CrossRef]
33. Ward, C.D.; Murchison, C.E.; Petroll, W.M.; Robertson, D.M. Evaluation of the Repeatability of the LacryDiag Ocular Surface Analyzer for Assessment of the Meibomian Glands and Tear Film. *Transl. Vis. Sci. Technol.* **2021**, *10*, 1. [CrossRef]
34. Lins, A. Image-based Eye Redness Using Standardized Ocular Surface Photography. In Proceedings of the 2nd MCI Medical Technologies Master's Conference, Innsbruck, Austria, 27–29 September 2021.
35. Gaudenzi, D.; Mori, T.; Crugliano, S.; Grasso, A.; Frontini, C.; Carducci, A.; Yadav, S.; Sgrulletta, R.; Schena, E.; Coassin, M.; et al. AS-OCT and Ocular Hygrometer as Innovative Tools in Dry Eye Disease Diagnosis. *Appl. Sci.* **2022**, *12*, 1647. [CrossRef]
36. Guillon, J.P. Abnormal lipid layers. Observation, differential diagnosis, and classification. *Adv. Exp. Med. Biol.* **1998**, *438*, 309–313.
37. Fodor, E.; Hagyó, K.; Resch, M.; Somodi, D.; Németh, J. Comparison of Tearscope-plus versus slit lamp measurements of inferior tear meniscus height in normal individuals. *Eur. J. Ophthalmol.* **2010**, *20*, 819–824. [CrossRef]
38. Markoulli, M.; Duong, T.B.; Lin, M.; Papas, E. Imaging the Tear Film: A Comparison Between the Subjective Keeler Tearscope-Plus™ and the Objective Oculus® Keratograph 5M and LipiView® Interferometer. *Curr. Eye Res.* **2018**, *43*, 155–162. [CrossRef]
39. Pult, H. Relationships between meibomian gland loss and age, sex, and dry eye. *Eye Contact Lens* **2018**, *44*, S318–S324. [CrossRef]
40. Ifrah, R.; Quevedo, L.; Gantz, L. Repeatability and reproducibility of Cobra HD fundus camera meibography in young adults with and without symptoms of dry eye. *Ophthalmic Physiol. Opt.* **2023**, *43*, 183–194. [CrossRef]
41. Coassin, M.; Mori, T.; Di Zazzo, A.; Poddi, M.; Sgrulletta, R.; Napolitano, P.; Bonini, S.; Orfeo, V.; Kohnen, T. Effect of minimonovision in bilateral implantation of a novel non-diffractive extended depth-of-focus intraocular lens: Defocus curves, visual outcomes, and quality of life. *Eur. J. Ophthalmol.* **2022**, *32*, 2942–2948. [CrossRef]
42. Bandello, F.; Coassin, M.; Di Zazzo, A.; Rizzo, S.; Biagini, I.; Pozdeyeva, N.; Sinitsyn, M.; Verzin, A.; De Rosa, P.; Calabrò, F.; et al. One week of levofloxacin plus dexamethasone eye drops for cataract surgery: An innovative and rational therapeutic strategy. *Eye* **2020**, *34*, 2112–2122. [CrossRef] [PubMed]
43. Antonini, M.; Gaudenzi, D.; Spelta, S.; Sborgia, G.; Poddi, M.; Micera, A.; Sgrulletta, R.; Coassin, M.; Di Zazzo, A. Ocular Surface Failure in Urban Syndrome. *J. Clin. Med.* **2021**, *10*, 3048. [CrossRef] [PubMed]

life

Review

Dynamic Aspects of Pre-Soft Contact Lens Tear Film and Their Relation to Dry Eye: Basic Science and Clinical Relevance

Norihiko Yokoi [1,*], Petar Eftimov [2] and Georgi As. Georgiev [3]

1 Department of Ophthalmology, Kyoto Prefectural University of Medicine, Kyoto 602-8566, Japan
2 Department of Cell and Developmental Biology, Faculty of Biology, Sofia University "St. Kliment Ohridski", 1164 Sofia, Bulgaria
3 Department of Optics and Spectroscopy, Faculty of Physics, Sofia University "St. Kliment Ohridski", 1164 Sofia, Bulgaria
* Correspondence: nyokoi@koto.kpu-m.ac.jp; Tel.: +81-75-251-5578

Abstract: Soft contact lens (SCL) perturbs the intimate connection between the pre-lens tear film (PLTF) and the ocular surface in various ways, i.e., (i) decrease in tear meniscus radius and aqueous tear thickness, (ii) attenuation of tear film lipid layer spread, (iii) limited wettability of SCL surface, (iv) increased friction with eyelid wiper, etc. This often results in SCL-related dry eye (SCLRDE) manifested as PLTF instability and contact lens discomfort (CLD). In this review, the individual contributions of factors (i–iv) to PLTF breakup patterns (BUP) and CLD are considered via the tear film-oriented diagnosis framework adopted by the Asia Dry Eye Society from a clinical and basic science perspective. It is shown that SCLRDE (due to aqueous deficiency, increased evaporation, or decreased wettability) and BUP of PLTF classify within the same types as the ones observed for the precorneal tear film. The analysis of PLTF dynamics reveals that the inclusion of SCL enhances the manifestation of BUP associated with (i) decreased thickness of PLTF aqueous layer and (ii) limited SCL wettability as shown by the rapid expansion of BUP area. PLTF thinness and instability result in increased blink-related friction and lid wiper epitheliopathy as major contributor to CLD.

Keywords: pre-lens tear film dynamics and stability; breakup patterns; soft contact lens; wettability; tear film lipid layer spread; dry eye

Citation: Yokoi, N.; Eftimov, P.; Georgiev, G.A. Dynamic Aspects of Pre-Soft Contact Lens Tear Film and Their Relation to Dry Eye: Basic Science and Clinical Relevance. *Life* **2023**, *13*, 859. https://doi.org/10.3390/life13040859

Academic Editors: Maria Pilar Vinardell, José-María Sánchez-González, Alejandro Cerviño and Carlos Rocha-de-Lossada

Received: 21 January 2023
Revised: 17 February 2023
Accepted: 16 March 2023
Published: 23 March 2023

1. Introduction

When a soft contact lens (SCL) is inserted at the ocular surface, the intimate relationship between the precorneal tear film (PCTF) and the corneal surface epithelium is perturbed. This may result in contact lens discomfort defined by the Tear Film & Ocular Surface Society workshop as "a condition characterized by episodic or persistent adverse ocular sensations related to lens wear either with or without visual disturbance, resulting from reduced compatibility between contact lens and the ocular environment, which can lead to decreased wearing time and discontinuation" [1]. It was concluded that one of the most important premises to solve SCL discomfort is to clarify the mechanisms behind the instability of pre-lens tear film (PLTF), which is also important to understand the pathophysiology of SCL-related dry eye (SCLRDE) [2].

In a SCL wearer, both tear menisci and precorneal tear film (PCTF) are split into pre- and post-lens parts [3]. Then, at eye opening, the tears at the inferior pre-lens tear meniscus (PLTM) are pulled up by the eyelid to deposit the PLTF. Thus, by modification of the structure and radius of the PLTM, SCL presence alters the very process of TF deposition and formation.

Once established, the PLTF often thins rapidly at 10–20 μm/min, far exceeding not only the PCTF thinning rates of 0.24–1.45 μm/min in healthy eyes but even the 7 μm/min observed for delipidated PCTF [4]. Therefore, it was long ago suggested that evaporation, although omnipresent at the ocular surface, cannot be the only factor behind PLTF dynamics.

Other factor thought to be crucially important is the difference in the wettability and durability of SCL surface compared to the corneal surface [4,5]. In contrast to the cornea whose wettability is maintained by membrane-associated mucins, especially the longest MUC16, SCL surface has intrinsic limitations as the contact lens lacks such glycocalyx coating. Indeed, while rabbit cornea maintains complete wetting (i.e., zero contact angle upon deposition of sessile water droplets) for ninety minutes after enucleation [6], soft contact lens displays finite contact angles within a broad range of 10° to 70° (depending on the SCL material and the measurement techniques used) even immediately after removal from the blister pack [7]. SCL wettability in the course of wear may be further diminished (i) by the in vivo deposition of proteins or lipids to the contact lens or (ii) by the increased susceptibility to adverse environmental conditions (low temperature, low humidity, etc.) due to the thinner aqueous layer of the PLTF (as compared to the PCTF) [8]. In contrast to the corneal surface, SCL surface has no regenerative property; it might be deteriorated to less wettable depending on the wearing time, which also leads to the association with decreased wettability [9].

Therefore, although SCLRDE is often considered as evaporative dry eye (EDE) [10], such a view does not reflect the plethora of perturbations that SCL wear exerts at the ocular surface. The Asia Dry Eye Society classifies dry eye (DE) into three types, viz. aqueous deficient DE (ADDE), increased evaporation DE (IEDE), and decreased wettability DE (DWDE), which allows to encompass the diverse challenges imposed on the tear film by SCL wear [11–14]. Thus, from the classification point of view, it is a matter of great interest what types of DE are associated with SCLRDE within the framework developed by the Asia Dry Eye Society.

Therefore, in this review, focusing on the pre-lens tear film dynamics, the mechanisms of the instability of PLTF and of SCLRDE are discussed from the perspective of basic science and clinical relevance. In the following sections, firstly (i) the clinical data on the effects of SCL wear on tear meniscus radius, tear film lipid layer (TFLL) spread, and PLTF breakup time and patterns will be presented (Section 2) and then, (ii) the theoretical perspective will be emphasized on how the SCL-induced perturbations of the ocular surface influence PLTF formation and stability (Section 3).

2. Structural and Functional Change of Tear Film in Soft Contact Lens Wear

The major goal of our contribution is to demonstrate the application of tear film-oriented diagnosis (TFOD) and tear film breakup patterns (BUP) classification as performed by the Asia Dry Eye Society [11–14] to the analysis of PLTF dynamics and (in)stability. The implementation of TFOD and BUP classification require comprehensive information about the PLTF dynamics, i.e., simultaneous knowledge about the SCL-induced perturbations of tear meniscus radius, TFLL spread grade, PLTF breakup time, and breakup pattern (i.e., size, location, and degree of expansion of the dry spot). We illustrate the abovementioned clinical phenomena with data (Sections 2.1 and 2.2) that were obtained by our team in the last decade [15–18]. The volunteers were healthy individuals without signs of dry eye or other ocular surface diseases, i.e., all the perturbations in PLTF dynamics can be unambiguously attributed to the insertion of SCL. These SCL-induced clinical phenomena at the ocular surface were reported also by other teams and key publications are referenced further. However, the data collected by our team are valuable considering the fact that the application of a broad range of mutually complimentary techniques allows to simultaneously and comprehensively evaluate how the different properties of PLTF change after SCL insertion and what happens prior to the fitting and shortly (30 min) after the removal of the SCL. Such observations are relatively rarely reported in the literature especially concerning the size, location, and degree and rate of expansion of breakup area(s), i.e., information important for the diagnostic algorithm of TFOD. Section 2.2 briefly summarizes the basics of the tear film-oriented diagnostics based on the analysis of tear film breakup patterns as developed in previous publications [11–14]. The relevance of TFOD to the analysis of PLTF dynamics is shown. Then, in Section 2.3 the importance of the increased friction realized between the SCL and the eyelid wiper due to the PLTF thinness and instability (i.e., PLTF provides less efficient lubrication than the "healthy" PCTF) is outlined.

2.1. Alteration of TMR, TFLL Spread and PLTF Stability in SCL Wear

In our SCL research, changes of tear volume, TFLL spread, and PLTF stability were evaluated noninvasively. Video-meniscometer [19,20] was employed for assessing tear meniscus radius (TMR). A video-interferometer, DR-1^{TM} (Kowa, Tokyo, Japan), was used [8] for the noninvasive evaluation of (i) the pattern and spread of TFLL (which both also qualitatively report about aqueous tear thickness [8] and (ii) the PLTF stability in terms of noninvasive breakup time (NIBUT) [8]. In our previous studies [15–18,21], it was found that after SCL insertion, (i) PLTM radius gets smaller than that of the original tear meniscus without SCL and (ii) that outside of the SCL, TMR is equivalent to the meniscus radius without SCL (Figure 1).

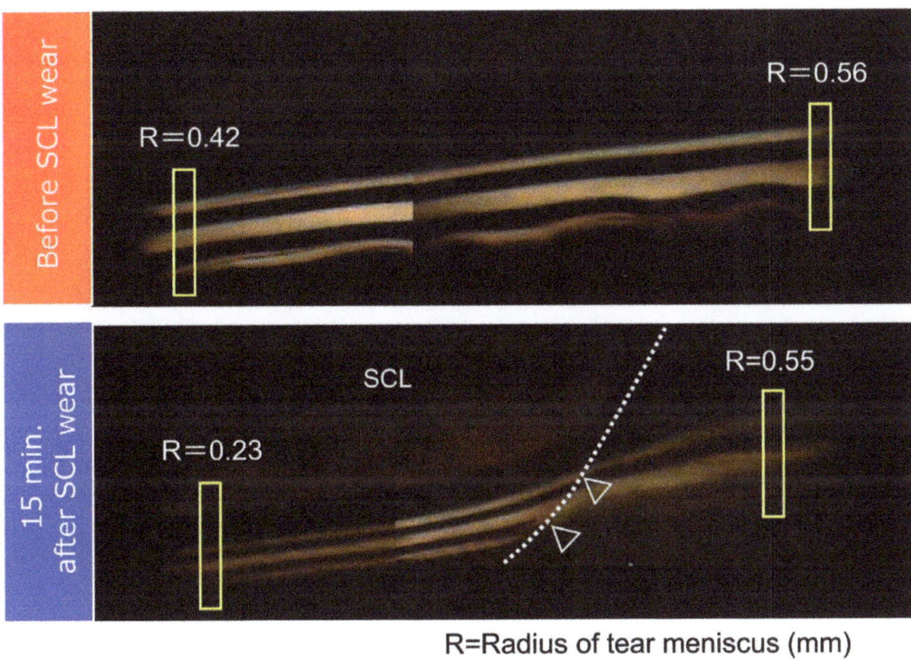

Figure 1. Tear meniscus change before and 15 min after SCL insertion. The upper image is the meniscometry image at the meniscus before SCL insertion, and lower image is that at 15 min after. Tear radius of tear meniscus (R) is attenuated only at the region where the SCL is inserted, which suggests that the meniscus tear volume is decreased only at the region. R: radius of tear meniscus (mm); SCL: soft contact lens. In the lower image the dashed line indicates imaginary margin of SCL behind the pre-SCL tear film and the arrow heads indicate the portion of the imaginary margin of SCL behind the pre-SCL tear meniscus.

These results show that TMR, an indicator for the meniscus tear volume [20], is decreased at the area where SCL is inserted. After the removal of the SCL, the original TMR is restored to the value prior to the SCL insertion, indicating that lowered TMR is a transient change observed only during the SCL wear (Figure 2). A similar pattern is observed regarding noninvasive breakup time (NIBUT), i.e., SCL insertion results in a temporary decrease in PLTF stability, which rapidly recovers to normal when SCL is removed (Figure 3).

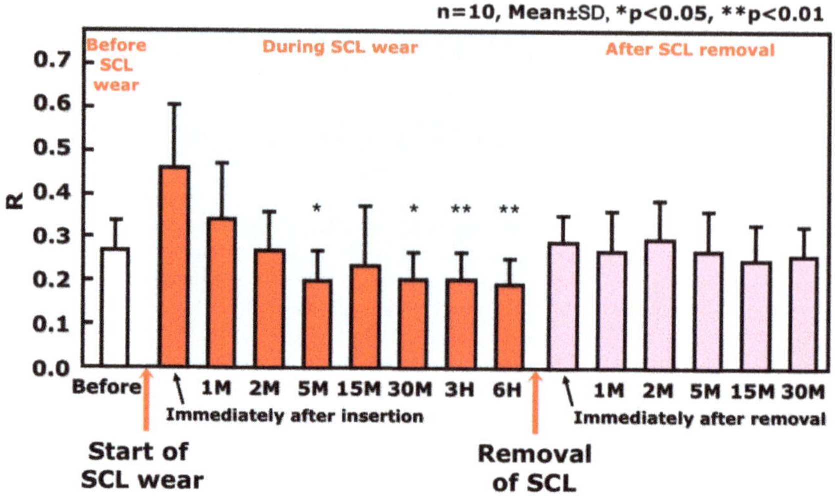

Figure 2. Time-Dependent Change in the tear meniscus radius (R), before, during, and after SCL wear R at the central lower tear meniscus corresponding to the meniscus tear volume, before, during, and after SCL wear. R is reduced as late as several minutes, say, 5 min after SCL insertion, maintained during wear, and is restored to that before SCL insertion immediately after SCL removal. R: radius of tear meniscus (mm); SCL: soft contact lens; M: minutes; H: hours. The figure is adapted with permission from reference [18] (Copyright year: 2015, copyright owner: The Japan Contact Lens Society).

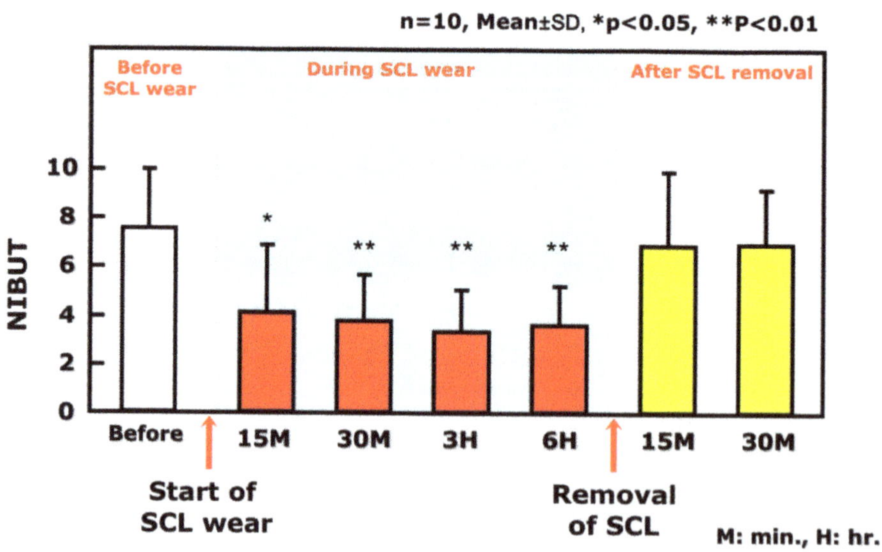

Figure 3. Time-Dependent Change in noninvasive breakup time (NIBUT), of the pre-lens tear film before, during, and after SCL wear. NIBUT corresponding to pre-LTF stability is reduced as late as 15 min after SCL insertion, is maintained during SCL wear, and is restored to that before SCL insertion immediately after SCL removal. SCL: soft contact lens; NIBUT: noninvasive breakup time (s); M: minutes; H: hours. The figure is adapted with permission from reference [18] (Copyright year: 2015, copyright owner: The Japan Contact Lens Society).

As demonstrated theoretically and clinically (see also Section 3) [22], there is a positive linear relationship between the radius of the lower tear meniscus and the aqueous layer thickness of PCTF [23–25]. Consequently, a similar relationship should be anticipated between the radius of the PLTM and the aqueous layer thickness of the PLTF. Hence, it is expected that in case of lower radius of the PLTM, PLTF thickness decreases, which in turn should manifest in (i) deteriorated grade of TFLL (Figure 4) and (ii) lower stability and shorter NIBUT of PLTF (Figure 5). All these phenomena were indeed observed clinically as shown in Figures 4 and 5 and are foreseen theoretically (see Section 3). PLTM radius is recovered after the removal of the SCL (Figure 2); therefore, the less stable PLTF during SCL wear is also a transient phenomenon, suggesting that the CL-related discomfort and SCLRDE are also transient symptoms and transient complications, respectively. In this regard, for the prevention of SCLRDE, an increase in aqueous tear volume is of essential importance, as it will result in increased tear film thickness and hence a raised PLTF stability.

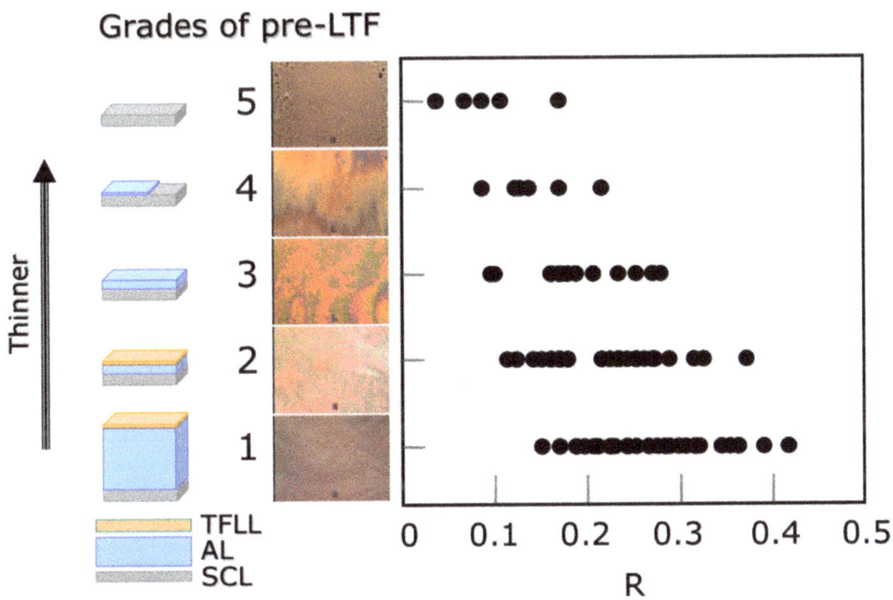

Figure 4. Relationship between R and interference grades of the pre-lens tear film. Positive correlation is seen between R and interference grades of pre-LTF ($r_s = -0.597$, $p < 0.0001$, n = 100), suggesting that an eye with lower tear volume is more susceptible to the thinner PLTF, which would lead to SCL-related dry eye. PLTF thickness is classified into one of 5 grades as shown here. In this classification, a greater grade corresponds to a thinner PLTF: Grade 1: Interference observed only from the tear film lipid layer (TFLL); Grade 2: Interference from the TFLL as well as a thin aqueous layer (AL); Grade 3: Interference from only a thin AL. TFLL is no longer present; Grade 4: Thin AL and the exposed surface of the SCL; Grade 5: No TF at all on the surface of the SCL. R: radius of tear meniscus (mm). One hundred eyes from 50 contact lens wearers were enrolled (60 eyes from 30 males; 40 eyes from 20 females; 32.2 (mean) ± 6.4 (SD) years old; with no dry eye) and they wore brand-new daily-use, one-day disposable soft contact lenses in the study. Measurement of radius of tear meniscus, which was followed by measurement of NIBUT, was performed 15 min after wearing of soft contact lens. The figure is adapted with permission from reference [18] (Copyright year: 2015, copyright owner: The Japan Contact Lens Society).

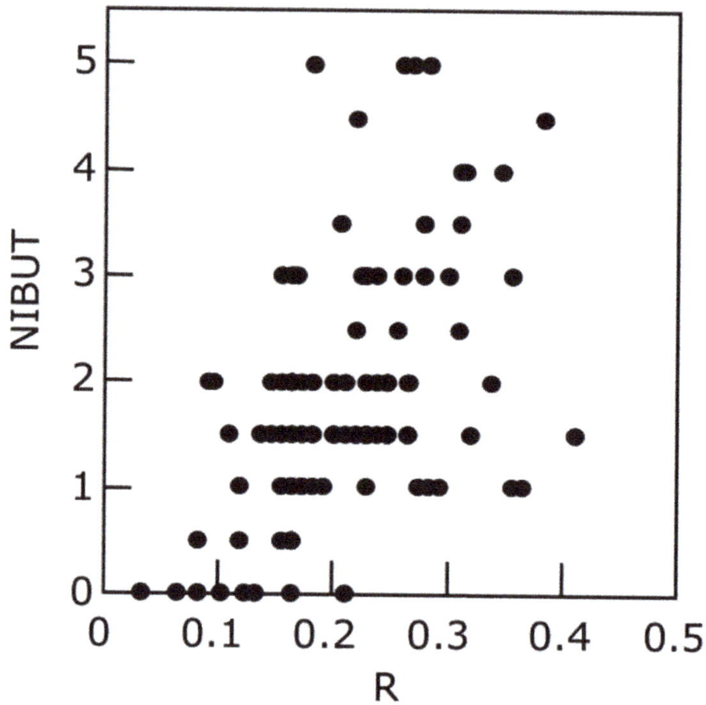

Figure 5. Relationship between R, the meniscus radius, and NIBUT. Positive correlation is seen between R and NIBUT (r_s = 0.476, $p < 0.0001$, n = 100), suggesting that an eye with lower tear volume is more susceptible to the instability of the pre-lens tear film which would lead to SCL-related dry eye. NIBUT: noninvasive breakup time (s); R: radius of tear meniscus (mm). One hundred eyes from 50 contact lens wearers were enrolled (60 eyes from 30 males; 40 eyes from 20 females; 32.2 (mean) ± 6.4 (SD) years old; with no dry eye) and they wore brand-new daily-use, one-day disposable soft contact lenses in the study. Measurement of radius of tear meniscus, which was followed by measurement of NIBUT, was performed 15 min after wearing of soft contact lens. The figure is adapted with permission from reference [18] (Copyright year: 2015, copyright owner: The Japan Contact Lens Society).

The SCL-induced decrease in tear meniscus radius and/or volume with accompanying drop in the PLTF thickness and noninvasive breakup time of healthy individuals is reported in multiple studies [3,26–28] including in 84 young (22.4 ± 2.6 years) healthy volunteers (i.e., a group characterized with stable and robust tear film) [29]. Particularly interesting in the context of dynamic changes in TMR is the study of Chen et al. [30] showing continuous decrease in the lower tear meniscus volume of healthy individuals from 0.6 μL to 0.48 μL (i.e., drop of TMR from 0.3 to 0.2 mm [20]) along 2 to 10 h of continuous SCL wear, which is in very good agreement with the magnitude of the effect observed by us.

The impaired spread of TFLL due to the thin PLTF (it is generally thought that PLTF is 1.5 μm thinner than the PCTF of a healthy individual) is also reported by multiple studies with various observational techniques such as Lipiview, Tearscope, video interferometry, etc. [31–34]. Guillon proposed a grading scale of the TFLL images obtained by tearscope which confirmed that in presence of SCL, the spread of TFLL is suppressed and the lipid layer structure becomes patchier and more heterogeneous [33,35], i.e., a result identical to that shown at Figure 4. The advantage of DR-1 observations is that compared to Lipiview or Tearscope evaluation, the visualizations are of higher resolution, contain more details, and encompass the entire ocular surface which facilitated the unambiguous interpretation of the results obtained.

2.2. Breakup Pattern of PLTF in SCL Wear

DE is manifested by unstable TF [10,36] and according to the Asia Dry Eye Society classification, different types of DE are characterized by different breakup patterns (BUPs) of the tear film identified both via invasive (fluorescein staining) and noninvasive (DR-1^{TM} video-interferometer) observations of PCTF [12,14,37]. As will be discussed further (here and in Section 3), the noninvasively observed BUPs are in good agreement with the theoretical and clinical knowledge of the dynamic changes that take place in the PCTF upon eye opening and at interblink. As these processes remain relevant, when CL is inserted at the ocular surface it can also be expected that identical BUPs will be observed in PLTM as well. At the SCL surface, just like at the corneal one, the TF establishment is realized in two steps involving (i) deposition of aqueous tears to the SCL surface at the upward movement of the upper eyelid and (ii) subsequent redistribution of the aqueous tears and of the tear film lipid layer across the PLTF. The processes involved in the PLTF formation together with the possible resultant BUPs as observed by DR-1^{TM} are introduced below in agreement with the previously developed classification of tear film breakup patterns [12,14,37] (Figures 6 and 7).

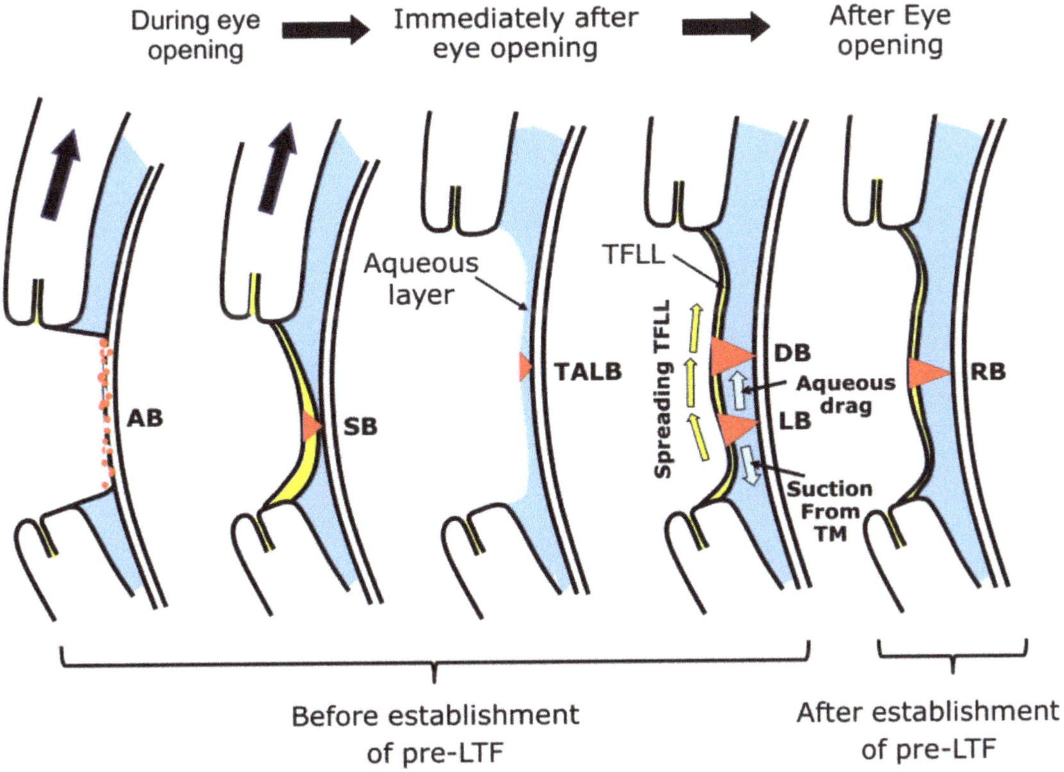

Figure 6. Classification of Breakup Patterns of PLTF with eye opening and with the eye kept open. Here yellow arrows imply upward spreading of TFLL and black arrows imply eyelid opening. See Sections 2.2 and 3 of the main manuscript for more detailed explanation of the events involved in pre-lens tear film dynamics at eye opening and at interblink.

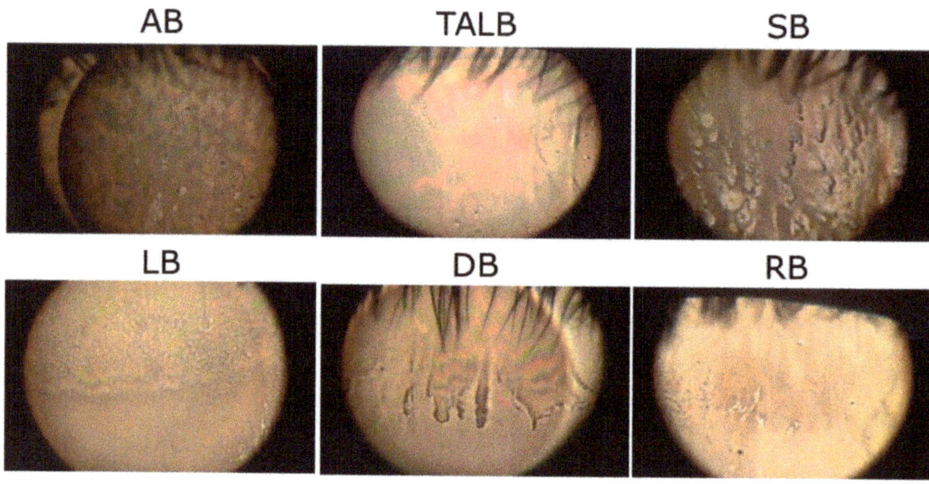

Figure 7. Classification of Breakup Patterns of Pre-LTF. AB: area break; TALB: thin aqueous layer break; SB: spot break; LB: line break; DB: dimple break; RB: random break.

(1) Deposition of aqueous tears to the SCL surface occurs at eye opening when the capillary suction pressure [38] from the upper PLTM pulls the aqueous tears up from the lower PLTM, which results in the coating of aqueous tears at the SCL surface. The efficiency of the deposition depends both on the meniscus radius and on the wettability of the SCL [9]. Thus, if the volume of PLTM is extremely diminished, there is no efficient deposition of aqueous tears on the surface of the SCL at all. This BUP corresponds to the "Area break" typical for severe aqueous tear deficient dry eye [21,22,37]. In slightly less severe cases of PLTM reduction, a limited amount of AT is deposited at the SCL surface to yield a PLTF that is too thin and does not allow for lipid layer spread (and the associated upward drag of aqueous tears) to proceed over it [39]. In this case, within 1 s, Thin-Aqueous Layer break occurs [37], which is also associated with the rapid expansion of the BUP due to the rapid drying and diminished wettability of the SCL surface. Sometimes, although rarely, the so-called "Spot break" (analogous to the spot break BUP of PCTF) can be observed at eye opening when the wettability of the SCL is locally impaired after extended wear due to the excessive deposition of lipids or proteins over a discrete region of the SCL, which results in rapid local rupture of the PLTF over that region.

(2) After eye is opened, and if a sufficiently thick aqueous tear film is established, upward spread of the TFLL occurs analogously to the dynamics of PCTF formation. The leading edge of the lipid layer spread drags the underlying aqueous tears upwards, which results in transient thinning of the aqueous layer downwards and its thickening upwards [39–41]. During this redistribution process of PLTF, two distinct BUPs of PLTF are differentiated, "Line break" and "Dimple break", which at this stage of tear film formation are also seen in PCTF, i.e., in dry eye over the native (in absence of SCL) corneal surface [12,14,37]. However, it should be noted that as the SCL surface is inherently less wettable than the corneal surface, both "Line break" and "Dimple break" of PLTF are accompanied by rapid expansion of the breakup area [12,14]. If PLTF remains stable until completion of the upward spread of the TFLL (i.e., in cases of higher number of aqueous tears at TMR $\geq$ 0.3 mm), after the cessation of TFLL upward spread, breakup appears as "Random break" corresponding to the eponymous BUP of the PCTF in DE without SCL wear. Random break may be associated with rapid expansion of the dry (i.e., breakup) area when the wettability of SCL surface is impaired.

Among those BUPs seen at the SCL surface, it may be suggested that "Area break" and "Thin aqueous layer break" are associated with relatively severe aqueous tear deficiency; "Spot break" is associated with the deterioration of the SCL surface; and Line break and Dimple break both with rapid expansion of the dry area are associated with decreased wettability of SCL surface. Random break occurs in cases with higher aqueous tear volume when thicker PLTF is formed. However, further investigation is necessary to clarify those BUPs. The importance of BUP dynamics as a tool to evaluate PLTF interactions with SCL and for differential diagnosis of dry eye is recently increasingly recognized as an important diagnostic modality in (video)keratography studies where there is ongoing research for the implementation of TFOD [42,43].

2.3. Comprehensive Understanding of the Vicious Cycle of SCLRDE

Due to the transient SCL-induced decrease in TMR, the PLTF is significantly thinned similarly to the condition observed in aqueous tear deficient PCTF. As SCL protects the cornea, the decreased PLTF stability should not result in corneal desiccation. However, contact lenses in general have a higher coefficient of friction than the human cornea, which deteriorates further with SCL dehydration at wear [44,45]. Therefore, the lack of thick and stable TF over the SCL should result in increased (higher than the one in precorneal tear film (PCTF)) blink-related friction and this mechanism should play a major role in the symptoms of contact lens-related discomfort and eye pain in SCLRDE (Figure 8).

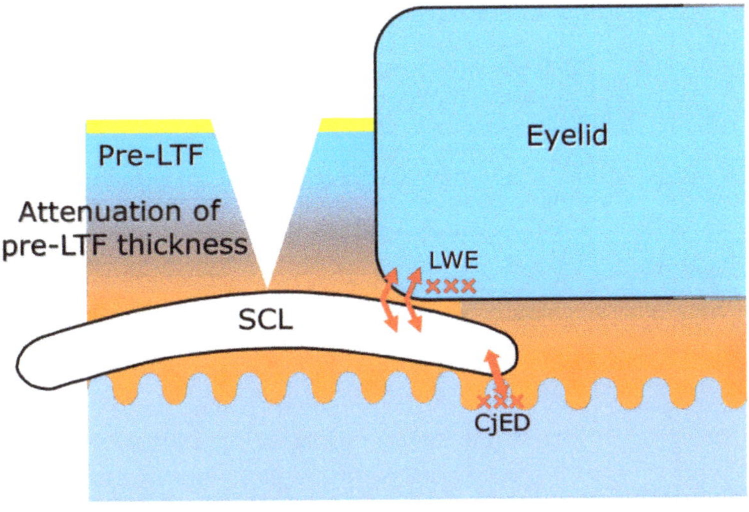

Figure 8. Possible mechanism for soft contact lens-related dry eye. When the lens is in place, the PLTF gets thin and unstable. As a result, blink-related friction occurs, leading to lid wiper epitheliopathy (LWE) and conjunctival epithelial damage (CjED), which in turn can cause contact lens discomfort (CLD). Here "x" implies the epithelial damage of the lid wiper portion or conjunctival epithelium and the arrows indicate the region of increased friction. In cases of aqueous tear deficiency, these mechanisms are more pronounced and the related CLD can be more severe.

Indeed, as described in an earlier review on the application of TFOD to SCL-related dry eye [46], lid wiper epitheliopathy (LWE) was found in 80% of SCLRDE subjects but in only 13% of asymptomatic contact lens wearers and thus an association is revealed between CLRDE and LWE [47]. In Japan, Shiraishi et al. [48] also reported a high frequency of LWE among contact lens wearers. A number of recent reviews highlighted LWE as a major factor associated with SCL dropout as well [49,50].

This was also shown in our previous study [18], where the radius of the PLTM decreases compared with the original PCTM within 15 min after SCL insertion (Figure 2), which results in thinner PLTF. This in turn leads to the increased friction between the SCL surface and the lid wiper portion (with associated epithelial damage at the lid wiper portion) [47] and between the edge of the SCL and the conjunctival surface [51]. Furthermore, eye dryness during SCL wear is found to be related to tear volume, conjunctival staining, and lid wiper epitheliopathy (LWE) scores [18,21]. However, no significant relationship was noted between corneal epithelial damage and eye dryness. Therefore, the SCL-covered corneal surface is protected against both desiccation and increased friction (as there is no direct contact with the lid wiper). Thus, the increased blink-related friction due to the thinned PLTF causes contact lens discomfort via two other mechanisms: (i) lid wiper epitheliopathy and (ii) conjunctival epithelial damage induced by the lens edge. It was found [15–18] that more than 3 h are needed for CLD to occur despite very early decrease in the PLTM radius and the attenuation of PLTF noninvasive breakup time only 30 min. after SCL insertion. Thus, commonly observed were cases with extreme dryness during SCL wear in which, despite minimal or no corneal epithelial staining after SCL removal, severe LWE and bulbar conjunctival damage were displayed. These results are in excellent agreement with previous reports that the severity of LWE [47] and the severity of bulbar conjunctival damage are associated with the severity of contact lens discomfort (CLD) [51]. Therefore, in SCL wear, it is not exaggerated to say that the problem of corneal surface damage due to the instability of PCTF is shifted to that of conjunctival surface damage because of increased blink-related friction due to the instability of PLTF.

Recently, greater attention has been paid from the point of view of material science to the enhancement of wettability and lubricity at the stage of SCL materials design [44]. To overcome SCLRDE and resultant CLD, SCLs with enhanced wettability and lubricity, and eye drops which can increase the PLTM thickness and/or decrease the friction, can be suggested as beneficial and further development has been advancing.

3. Some Fundamental Aspects of PLTF Dynamics

Already in 2005, Nichols et al. [4] mentioned that although SCLRDE is typically classified as evaporative dry eye due to the accelerated PLTF thinning rate, this is an oversimplification as CL presence may exert a variety of perturbations at the ocular surface. In line with these thoughts, the current study shows that CL may reduce tear meniscus radius, inhibit TFLL spread, or cause rapid PLTF breakup that might be more related to the CL wettability rather than to evaporation or to aqueous tear (AT) deficiency. All these factors, which were indeed found to take place at the ocular surface, are expected to reduce the stability of PLTF in a specific manner distinct from the evaporation and will be discussed in the next subsections.

3.1. Pre-lens Tear Meniscus Radius, Aqueous Tear Film Thickness, and TFLL Spread

In the initial step of tear film formation, the upper eyelid deposits the aqueous tear across the ocular surface at upstroke similarly to the painting of a wall with a brush. The deposition efficiency is determined by (i) the tear meniscus radius (R, mm) and surface tension (γ = 43 mN/m) which set the capillary action of the meniscus, (ii) the eyelid velocity (U, cm/s) in the course of the upstroke, and (iii) the AT viscosity (μ = 0.013 cP) responsible for the viscous drag opposing to the deposition [52,53]. Furthermore, R determines the tear meniscus volume, which is a measure for the overall amount of AT present at the ocular surface [20].

Assuming a perfectly extensible lipid layer, the relationship between these parameters and the initial thickness (h) of the TF at deposition is given by Equation (1) [22]:

$$h = 1.338R(\mu\frac{U}{\gamma})^{2/3} \quad (1)$$

The spatiotemporal eyelid velocity distribution [52,53] is known from the literature (Figure 9, left panel) and the upper and lower tear meniscus radius and the CL influence on them are known to be similar [54]. Thus, it is possible to quantitatively evaluate the impact of the reduction of tear meniscus radius caused by CL insertion as found here and in previous studies (Figure 9, right panel).

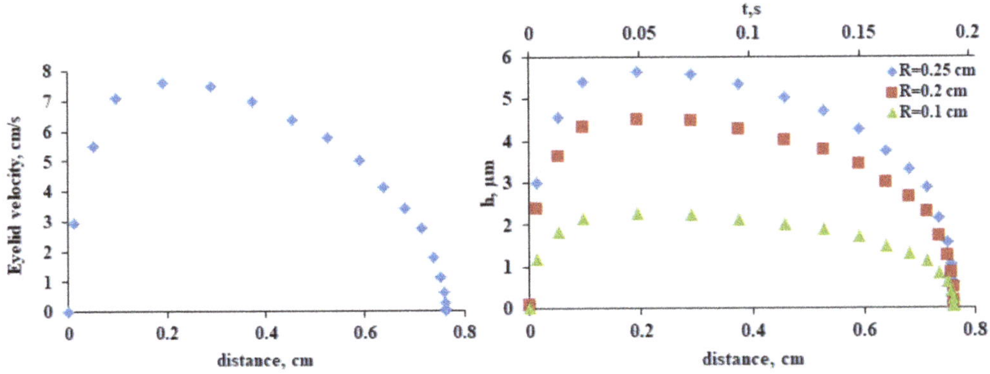

Figure 9. Left panel: The velocity of upper eyelid across the ocular surface. The eyelid traverses the distance for 0.2 s. **Right panel**: Thickness profile of PLTF immediately after deposition as calculated by Equation (1). At both panels, $x = 0$ is positioned at the lower eyelid margin.

It should be kept in mind that the PCTF thickness profile calculated with Equation (1) represents h values instantaneously at deposition, i.e., prior TFLL upward spread and black line formation to take place, which results in 1–1.5 µm thinning of TF over the central cornea. Thus, these thickness profiles are somewhat different from the commonly measured stationary shapes reported in literature. Still, the calculated TF thicknesses are well within the range of TF thickness values measured with high resolution optical coherence tomography (OCT) [55].

As can be seen, the decrease in tear meniscus radius typical in the presence of CL reduces the thickness of the AT deposited at eye opening, similarly to the complications arising in the case of ADDE. The theoretical result is in excellent agreement with the clinical data here as well as with previous studies on SCL-related dry eye [56,57]. It was reported that CL wear reduces by one-third the tear meniscus volume, which is associated with a decrease in the total tear volume and of the AT thickness, which in turn correlates with the degree of CL discomfort and with accelerated destabilization of the PLTF [42,58].

The significantly decreased thickness of the AT layer upon deposition not only facilitates rapid breakup by its own self but it also prevents the next crucial step of the TF formation: the upward spread of TFLL from the lower to the upper tear meniscus [59–62]. This is so because in order for TFLL spread to take place, a sufficiently thick AT layer is obligatory in order to prevent friction of TFLL with the CL surface [39]. In turn, TFLL spread drags the underlying AT upwards, which results in redistribution of the AT (i.e., thinning of the TF at the lower and central part of the ocular surface and TF thickening at the upper cornea) to more uniform TF thickness profile across the ocular surface [39–41].

The rheology of TFLL spread is well described by the Voigt law (Equations (2) and (3)) [59]:

$$H(t) = H_{max}[1 - \exp(-\frac{t}{\tau})], \quad (2)$$

$$H_{max} = V_0 \tau \quad (3)$$

Here, H is the distance (the height) travelled by the TFLL at time t, H_{max} is the maximum height reached when TFLL spread is complete, V_0 is the initial TFLL velocity at $t = 0$, τ is characteristic time defined by the ratio of TFLL elasticity to the friction experienced by TFLL upon spread, and t is time.

Analysis of clinical data revealed strong and statistically significant correlations between the rheological and the clinical parameters, which revealed that with the increase in AT thickness (and of tear meniscus radius, respectively) the velocity, the characteristic spread time, and the maximum height travelled by the TFLL are enhanced: $V_0 \approx 0.43718 \times h^2$ (R = 0.75; p = 0.017) and $\tau \approx 1.7211 \times h^{-1}$ (R = 0.64; p = 0.014) (in these empirical correlations, t is in seconds, V_0 in mm/s, and h in micrometers) [59–62].

This semiempirical framework allows to elucidate the typical impact of the decrease in R (i.e., of AT thickness) on the TFLL spread (Figure 10).

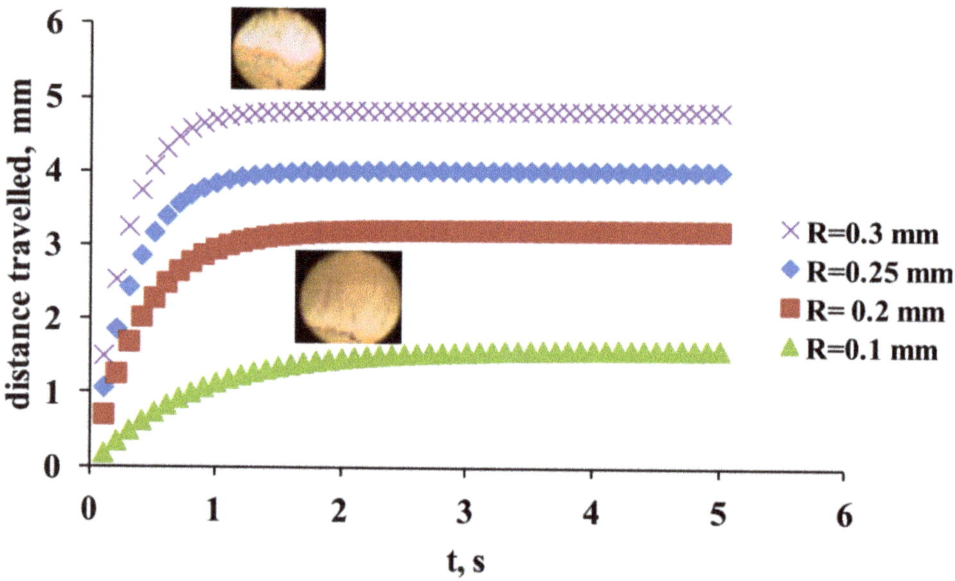

Figure 10. Higher tear meniscus radius means higher thickness of AT layer at deposition and faster and better spread of TFLL (Equations (2) and (3)), as illustrated by the image inserts. The calculated curves are in very good agreement with the trends of clinical data.

As can be seen, the reduction in tear meniscus radius (i.e., of AT volume across the ocular surface) results in slower and incomplete TFLL spread, which in turn prevents the formation of proper tear film [59]. These estimates are in very good agreement with the clinical results reported previously [59–62]. Thus it can be seen that the SCLRDE in which tear meniscus radius is decreased is more similar in its mechanism to ADDE, rather than to evaporative dry eye [14]. In addition, the CL-related reduction of tear meniscus radius is strongly correlated with certain breakup patterns (primarily "Area break", "Thin aqueous layer break", and "Line break") which are typical for severe and mild to moderate aqueous tear deficiency, respectively, when precorneal tear film instability is concerned [14]. An important difference between these forms of SCLRDE and typical ADDE is that in the case of SCLRDE, the reduction of tear meniscus radius and of AT thickness at deposition is often not due to the inherent AT deficiency due to lacrimal gland dysfunction but due to the perturbation of tear meniscus structure by the contact lens. Thus, once the SCL is removed from the eye, the tear meniscus curvature is restored to normal and the stability of the precorneal tear film returns to normal as well [42,58].

3.2. CLs and TF Thinning Rate in the Region of the Blackline

Another very important phenomenon in healthy and "dry" eyes related to the capillary action of tear menisci is the formation of "black line", i.e., dark (when visualized with fluorescein) thin ($\leq$200 nm) lines near the upper and lower lid margins, immediately following a blink [38] and resulting in the formation of a "perched tear film" over the central cornea. Although TF may not break up at black lines, the faster their formation is, the higher is the suction of aqueous tears towards the tear menisci and the faster the thinning of the adjacent TF regions. This in turn promotes TF instability starting over the lower towards the central cornea [63].

Numerical solutions of the thin film equation allowed Sharma et al., 1998 [64] to derive a hydrodynamic model (Equation (4)) for the meniscus-induced thinning of the tear film:

$$t = 3.52\left(\frac{R}{h_m}\right)^{2.232}\left(\frac{h_0}{R}\right)^{1.268}\left(\mu\frac{RC}{\gamma}\right) \tag{4}$$

Here, t is TF thinning time, R is tear meniscus radius, h_0 and h_m are the initial and minimal TF thickness, respectively, μ is tear viscosity, and γ is tear surface tension. Equation (4) also contains the term C which accounts for the capability of the TFLL to provide tangentially immobile air/tear surface (Gibbs-Marangoni effect). $C = 1$ for a "free" tear surface of zero shear stress, whereas $C = 4$ when the tear film surface is immobilized by the lipid layer. A very similar solution of the thin film equation for the meniscus-induced TF thinning was also derived by Wong et al., 1996 [52].

Equation (4) allows to model the impact of CL-induced decrease in tear meniscus radius on the rate of meniscus-induced thinning (Figure 11).

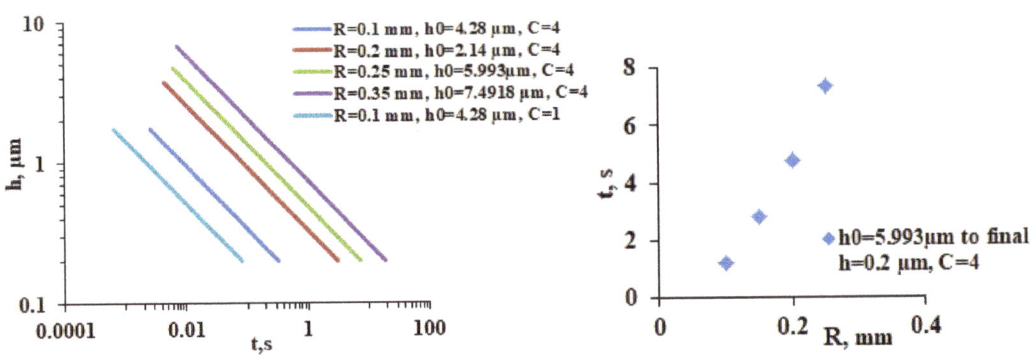

Figure 11. Left panel: lower tear meniscus radius is associated with lower AT thickness and faster thinning as calculated by Equation (4). The final thickness at the black line is 0.2 µm. **Right panel**: Illustration of the "thirsty meniscus phenomenon": at equal other conditions, lower tear meniscus radius has stronger suction capillary pressure and greatly accelerates film thinning.

As can be seen in Figure 11, left panel, the decrease in tear meniscus radius and the associated decrease in PLTF thickness result in faster thinning rates. The effect can be additionally emphasized if dysfunctional TFLL ($C = 1$) is included, a condition that is well possible in the case of a very thin AT layer (i.e., very low R), where TFLL spread is suppressed. The effect of tear meniscus radius at equal other conditions, i.e., equal AT thickness and TFLL spread (which is of course physiologically unrealistic) is simulated at Figure 11, right panel. It shows that the lower is the tear meniscus radius, the faster is the PLTF thinning. This is a demonstration of the so-called "thirsty" meniscus, i.e., lower R means higher tear meniscus curvature and stronger suction pressure exerted by the meniscus at the adjacent tear film. Such a trend is confirmed in clinical observation and is known to result in accelerated breakup of the TF at the lower cornea in ADDE [63]. Such behavior corresponds to the "Area break" and "Thin aqueous layer break" BUPs commonly observed over CL surface at low R values. Once again, this mechanism of accelerated instability of PLTF is related not with evaporation but with the higher suction pressure and AT deficiency at deposition induced by the more curved tear meniscus.

3.3. How CL Wettability May Alter PLTF Thinning Rate and Stability Away from the Black Line Region

In vivo studies revealed that even if successfully deposited (i.e., the CL-induced reduction of R is moderate), PLTF often has an abnormally high thinning rate of 10–20 μm/min compared to the "standard" PCTF thinning rates of 0.24–1.45 μm/min and even faster than the 7 μm/min observed for delipidated PCTF [4,65–67]. Such abnormally accelerated thinning cannot be explained solely by evaporation and is commonly attributed to dewetting [68]. This indeed can be expected, as most CLs display significant water contact angles (θ = 30° to ≥60°) immediately after removal from blister which tend to increase during wear [7,9]. In contrast, freshly enucleated rabbit cornea with intact glycocalyx displayed ideal wettability (θ = 0°) for 90 min after exposure to air [6].

In the eye, PLTF is continuously subjected to perturbations due to eyelid and eye movements, invasion by dust/cosmetic particles, and formation of hydrophobic nonwetting spots at the SiHy surface (due to silicone migration, lipid deposition, attachment of tear micro bubbles or other), etc. [64]. These perturbations may open transient holes of micrometric size in it, which might either (i) enclose, i.e., the PLTF is repaired and remains stable, or (ii) expand and result in dewetting of the SiHy and destabilization of the PLTF. Sharma and Ruckenstein [69,70] derived the relationship between the PLTF thickness and stability, the hole radius, and the water contact angle (i.e., the wettability) of the SiHy (Equation (5)):

$$\left(\frac{h_c}{r}\right) = tan\ \theta\left[\sqrt{1+cos\theta}-1\right],\ for\ \theta \leq \frac{\pi}{2} \quad (5)$$

Here, h_c is the critical thickness below which the wetting film is no longer unconditionally stable and becomes susceptible to rupture if a hole, a dry spot, with radius r, is formed on the SiHy surface. For CA ≤ 45°, the h_c/r ratio is <0.4, i.e., even thin (compared to the size of the defect) PLTF should remain stable. This means that for a nonwetting spot of 3–4 micrometer radius (as typically observed at the early stages of SiHy desiccation damage) [71], for any aqueous film thickness ≥1.6 μm, the wetting film will remain unconditionally stable. With the raise of h_c/r due to macroscopic increase of the contact angle or due to locally compromised CL surface (i.e., due to lipid deposition, microbubble attachment, etc.), the situation changes. For an h_c/r ratio of ≥0.5, the PLTF will become susceptible to rupture for h < 2 μm. This means that at the typical physiological PLTF thickness of 2–2.5 μm, dewetting becomes energetically favorable, which agrees very well with the rapid thinning and instability of PLTF commonly observed in the current study and elsewhere [39–41].

A critical moment in the TF formation that can promote the contamination of the CL surface with lipids/particles is the TFLL upward spread in which a Marangoni ridge is formed, i.e., thick AT is dragged upwards by lipid layer while AT thickness behind the advancing front is temporarily depressed which may result in TF destabilization over that region [39–41]. This redistribution of tear fluid is registered as up to 1.5 μm AT thinning over the central cornea within 1 s after eye opening [39–41]. This mechanism is thought to be operative for the line break and dimple break registered within 1–2 s at the CL regions located over the lower or central part of the cornea [12].

Apart from facilitating TF rupture at micrometric thicknesses, higher contact angle of the contact lens is also responsible for higher rate and degree of expansion of the dry area over the CL surface, which additionally enhances PLTF instability with the accumulation of desiccation stress during daily wear. A detailed review of the possible relationship between the rate and degree of expansion of dry spots and the biomaterial contact angles can be found in Bertand et al., 2010 [72].

The agreement between the basic science considerations of CL wettability and the in vivo observations of PLTF (in)stability patterns in the current study align with the clinical correlations between higher CL wettability (lower water contact angle) and higher comfort score and PLTF breakup times for a broad range of CL materials [73–77].

3.4. The Random Break Pattern

When all the steps of PLTF formation are completed and uniform tear film is established at the SCL surface, tear film breakup occurs at a random position within 5–10 s over the CLs. This type of breakup is promoted by evaporation, alone or in combination with other factors discussed above, and resembles the random break observed in healthy eyes without CL.

Overall, it can be seen that the fitting of CL perturbs the ocular surface in multiple ways: reduction of the tear meniscus radius, inhibition of TFLL spread, and substitution of the highly wettable cornea with CL surface of limited wettability. All these perturbations may destabilize the PLTF in a unique way distinct from evaporation. This is indeed confirmed by the diverse type of PLTF breakup patterns that matched very well the PCTF instability patterns defined in the tear film-oriented diagnosis of dry eye developed for PCTF treatment [14,37].

4. Conclusions

In conclusion, this review discussed the mechanism of SCLDE from a clinical and basic science point of view. This scientific field needs interaction among clinicians, basic scientists, and SCL makers, and as far as SCL is important for the correction of vision, further advances are required to improve the safety of SCL to OS and decrease CLD.

Author Contributions: Conceptualization, N.Y. and G.A.G.; methodology, N.Y. and G.A.G.; validation, N.Y. and G.A.G.; formal analysis, N.Y., G.A.G. and P.E.; investigation, N.Y. and G.A.G.; data curation, N.Y. and G.A.G.; writing—original draft preparation, N.Y., G.A.G. and P.E.; writing—review and editing, N.Y., G.A.G. and P.E.; visualization, N.Y., G.A.G. and P.E. supervision, N.Y. and G.A.G. All authors have read and agreed to the published version of the manuscript.

Funding: This research was supported by JSPS KAKENHI for Scientific Research from the Japanese Ministry of Education, Culture, Sports, Science and Technology (Grant number: JP20K09794).

Data Availability Statement: The data that support the findings of this study are available from the corresponding author, N.Y., upon reasonable request and also in the references quoted in the text.

Conflicts of Interest: The authors declare no conflict of interest. The funders had no role in the design of the study; in the collection, analyses, or interpretation of data; in the writing of the manuscript; or in the decision to publish the results.

References

1. Nichols, K.K.; Redfern, R.L.; Jacob, J.T.; Nelson, J.D.; Fonn, D.; Forstot, S.L.; Huang, J.-F.; Holden, B.A.; Nichols, J.J. The TFOS International Workshop on Contact Lens Discomfort: Report of the Definition and Classification Subcommittee. *Investig. Ophthalmol. Vis. Sci.* **2013**, *54*, TFOS14–TFOS19. [CrossRef]
2. Craig, J.P.; Willcox, M.D.; Argüeso, P.; Maissa, C.; Stahl, U.; Tomlinson, A.; Wang, J.; Yokoi, N.; Stapleton, F. The TFOS International Workshop on Contact Lens Discomfort: Report of the contact lens interactions with the tear film subcommittee. *Investig. Ophthalmol. Vis. Sci.* **2013**, *54*, TFOS123–TFOS156. [CrossRef]
3. Nagahara, Y.; Koh, S.; Maeda, N.; Nishida, K.; Watanabe, H. Prominent Decrease of Tear Meniscus Height With Contact Lens Wear and Efficacy of Eye Drop Instillation. *Eye Contact Lens* **2015**, *41*, 318–322. [CrossRef]
4. Nichols, J.J.; Mitchell, G.L.; King-Smith, P.E. Thinning rate of the precorneal and prelens tear films. *Investig. Ophthalmol. Vis. Sci.* **2005**, *46*, 2353–2361. [CrossRef] [PubMed]
5. Szczesna-Iskander, D.H.; Iskander, D.R.; Alonso-Caneiro, D. Contact lens poor wettability and visual performance. *Contact Lens Anterior Eye* **2018**, *41*, S13. [CrossRef]
6. Tiffany, J.M. Measurement of wettability of the corneal epithelium. II. Contact angle method. *Acta Ophthalmol.* **1990**, *68*, 182–187. [CrossRef] [PubMed]
7. Keir, N.; Jones, L. Wettability and Silicone Hydrogel Lenses: A Review. *Eye Contact Lens* **2013**, *39*, 100–108. [CrossRef]
8. Maruyama, K.; Yokoi, N.; Takamata, A.; Kinoshita, S. Effect of environmental conditions on tear dynamics in soft contact lens wearers. *Investig. Ophthalmol. Vis. Sci.* **2004**, *45*, 2563–2568. [CrossRef]
9. Eftimov, P.; Yokoi, N.; Peev, N.; Georgiev, G.A. Impact of Air Exposure Time on the Water Contact Angles of Daily Disposable Silicone Hydrogels. *Int. J. Mol. Sci.* **2019**, *20*, 1313. [CrossRef]
10. Wolffsohn, J.S.; Arita, R.; Chalmers, R.; Djalilian, A.; Dogru, M.; Dumbleton, K.; Gupta, P.K.; Karpecki, P.; Lazreg, S.; Pult, H.; et al. TFOS DEWS II Diagnostic Methodology report. *Ocul. Surf.* **2017**, *15*, 539–574. [CrossRef]

11. Yokoi, N.; Georgiev, G.A.; Kato, H.; Komuro, A.; Sonomura, Y.; Sotozono, C.; Tsubota, K.; Kinoshita, S. Classification of Fluorescein Breakup Patterns: A Novel Method of Differential Diagnosis for Dry Eye. *Am. J. Ophthalmol.* **2017**, *180*, 72–85. [CrossRef] [PubMed]
12. Yokoi, N.; Georgiev, G.A. Tear Film-Oriented Diagnosis and Tear Film-Oriented Therapy for Dry Eye Based on Tear Film Dynamics. *Investig. Ophthalmol. Vis. Sci.* **2018**, *59*, DES13–DES22. [CrossRef]
13. Tsubota, K.; Yokoi, N.; Shimazaki, J.; Watanabe, H.; Dogru, M.; Yamada, M.; Kinoshita, S.; Kim, H.M.; Tchah, H.W.; Hyon, J.Y.; et al. New Perspectives on Dry Eye Definition and Diagnosis: A Consensus Report by the Asia Dry Eye Society. *Ocul. Surf.* **2017**, *15*, 65–76. [CrossRef]
14. Yokoi, N.; Georgiev, G.A. Tear-film-oriented diagnosis for dry eye. *Jpn. J. Ophthalmol.* **2019**, *63*, 127–136. [CrossRef]
15. Yokoi, N.; Sakai, R.; Yamaguchi, H.; Kinoshita, S. Effects of the peripheral thickness of soft contact lenses on tears and eye dryness during soft contact lens wear. *Investig. Ophthalmol. Vis. Sci.* **2016**, *57*, 1470.
16. Maruyama, K.; Yokoi, N.; Kinoshita, S. Relationship Between Dryness and Tear Dynamics During Soft Contact Lens Wear. *Investig. Ophthalmol. Vis. Sci.* **2006**, *47*, 84.
17. Yokoi, N.; Sakai, R.; Georgiev, G.A.; Kato, H.; Sotozono, C.; Kinoshita, S. Impact of Soft Contact Lens Wear on Tear Film Breakup Patterns, Meniscus Tear Volume, and Tear Film Stability. *Investig. Ophthalmol. Vis. Sci.* **2018**, *59*, 1751.
18. Yokoi, N. Contact-lens-related dry eye, tears and tear film. *J. Jpn. CL Soc.* **2015**, *57*, 222–235. (In Japanese)
19. Yokoi, N.; Bron, A.J.; Tiffany, J.M.; Kinoshita, S. Reflective meniscometry: A new field of dry eye assessment. *Cornea* **2000**, *19* (Suppl. 3), S37–S43. [CrossRef]
20. Yokoi, N.; Bron, A.J.; Tiffany, J.M.; Maruyama, K.; Komuro, A.; Kinoshita, S. Relationship between tear volume and tear meniscus curvature. *Arch. Ophthalmol.* **2004**, *122*, 1265–1269. [CrossRef]
21. Yamaguchi, M.; Nishijima, T.; Shimazaki, J.; Takamura, E.; Yokoi, N.; Watanabe, H.; Ohashi, Y. Real-world assessment of diquafosol in dry eye patients with risk factors such as contact lens, meibomian gland dysfunction, and conjunctivochalasis: Subgroup analysis from a prospective observational study. *Clin. Ophthalmol.* **2015**, *9*, 2251–2256. [PubMed]
22. King-Smith, P.E.; Fink, B.A.; Hill, R.M.; Koelling, K.W.; Tiffany, J.M. The thickness of the tear film. *Curr. Eye Res.* **2004**, *29*, 357–368. [CrossRef]
23. Creech, J.L.; Do, L.T.; Fatt, I.; Radke, C.J. In vivo tear-film thickness determination and implications for tear-film stability. *Curr. Eye Res.* **1998**, *17*, 1058–1066. [CrossRef] [PubMed]
24. Hosaka, E.; Kawamorita, T.; Ogasawara, Y.; Nakayama, N.; Uozato, H.; Shimizu, K.; Dogru, M.; Tsubota, K.; Goto, E. Interferometry in the evaluation of precorneal tear film thickness in dry eye. *Am. J. Ophthalmol.* **2011**, *151*, 18–23.e1. [CrossRef]
25. Stegmann, H.; Aranha Dos Santos, V.; Messner, A.; Unterhuber, A.; Schmidl, D.; Garhöfer, G.; Schmetterer, L.; Werkmeister, R.M. Automatic assessment of tear film and tear meniscus parameters in healthy subjects using ultrahigh-resolution optical coherence tomography. *Biomed. Opt. Express* **2019**, *10*, 2744–2756. [CrossRef] [PubMed]
26. Chen, Q.; Wang, J.; Shen, M.; Cai, C.; Li, J.; Cui, L.; Qu, J.; Lu, F. Lower volumes of tear menisci in contact lens wearers with dry eye symptoms. *Investig. Ophthalmol. Vis. Sci.* **2009**, *50*, 3159–3163. [CrossRef]
27. Young, G.; Chalmers, R.; Napier, L.; Kern, J.; Hunt, C.; Dumbleton, K. Soft contact lens-related dryness with and without clinical signs. *Optom. Vis. Sci. Off. Publ. Am. Acad. Optom.* **2012**, *89*, 1125–1132. [CrossRef]
28. Itokawa, T.; Okajima, Y.; Suzuki, T.; Kakisu, K.; Iwashita, H.; Murakami, Y.; Hori, Y. Association Between Ocular Surface Temperature and Tear Film Stability in Soft Contact Lens Wearers. *Investig. Ophthalmol. Vis. Sci.* **2018**, *59*, 771–775. [CrossRef]
29. Vicente García-Marqués, J.; Talens-Estarelles, C.; García-Lázaro, S.; Cerviño, A. Assessment of condition-induced changes on the ocular surface using novel methods to assess the tear film dynamics and the lipid layer. *Contact Lens Anterior Eye J. Br. Contact Lens Assoc.* **2022**, 101799. [CrossRef]
30. Chen, Q.; Wang, J.; Shen, M.; Cui, L.; Cai, C.; Li, M.; Li, K.; Lu, F. Tear menisci and ocular discomfort during daily contact lens wear in symptomatic wearers. *Investig. Ophthalmol. Vis. Sci.* **2011**, *52*, 2175–2180. [CrossRef]
31. Kaido, M.; Kawashima, M.; Ishida, R.; Tsubota, K. Tear Film Dynamics of Soft Contact Lens-Induced Dry Eye. *Curr. Eye Res.* **2020**, *45*, 782–788. [CrossRef] [PubMed]
32. Rohit, A.; Willcox, M.; Stapleton, F. Tear lipid layer and contact lens comfort: A review. *Eye Contact Lens* **2013**, *39*, 247–253. [CrossRef] [PubMed]
33. Guillon, M.; Styles, E.; Guillon, J.P.; Maïssa, C. Preocular tear film characteristics of nonwearers and soft contact lens wearers. *Optom. Vis. Sci. Off. Publ. Am. Acad. Optom.* **1997**, *74*, 273–279. [CrossRef] [PubMed]
34. Kopf, M.; Yi, F.; Iskander, D.R.; Collins, M.J.; Shaw, A.J.; Straker, B. Tear Film Surface Quality with Soft Contact Lenses Using Dynamic Videokeratoscopy. *J. Optom.* **2008**, *1*, 14–21. [CrossRef]
35. Brahim, I.; Lamard, M.; Benyoussef, A.-A.; Quellec, G. Automation of dry eye disease quantitative assessment: A review. *Clin. Exp. Ophthalmol.* **2022**, *50*, 653–666. [CrossRef] [PubMed]
36. Tsubota, K.; Yokoi, N.; Watanabe, H.; Dogru, M.; Kojima, T.; Yamada, M.; Kinoshita, S.; Kim, H.M.; Tchah, H.W.; Hyon, J.Y.; et al. A New Perspective on Dry Eye Classification: Proposal by the Asia Dry Eye Society. *Eye Contact Lens* **2020**, *46* (Suppl. 1), S2–S13. [CrossRef] [PubMed]
37. Yokoi, N.; Georgiev, G.A. Tear-film-oriented diagnosis and therapy for dry eye. In *Dry Eye Syndrome: Basic and Clinical Perspectives*; Yokoi, N., Ed.; Future Medicine Ltd.: London, UK, 2013; pp. 96–108. [CrossRef]
38. McDonald, J.E.; Brubaker, S. Meniscus-induced thinning of tear films. *Am. J. Ophthalmol.* **1971**, *72*, 139–146. [CrossRef]

39. Berger, R.E.; Corrsin, S. A surface tension gradient mechanism for driving the pre-corneal tear film after a blink. *J. Biomech.* **1974**, *7*, 225–238. [CrossRef]
40. Benedetto, D.A.; Clinch, T.E.; Laibson, P.R. In vivo observation of tear dynamics using fluorophotometry. *Arch. Ophthalmol.* **1984**, *102*, 410–412. [CrossRef]
41. Goto, E.; Dogru, M.; Kojima, T.; Tsubota, K. Computer-synthesis of an interference color chart of human tear lipid layer, by a colorimetric approach. *Investig. Ophthalmol. Vis. Sci.* **2003**, *44*, 4693–4697. [CrossRef]
42. Downie, L.E.; Craig, J.P. Tear film evaluation and management in soft contact lens wear: A systematic approach. *Clin. Exp. Optom.* **2017**, *100*, 438–458. [CrossRef]
43. Yokoi, N.; Kusada, N.; Kato, H.; Furusawa, Y.; Sotozono, C.; Georgiev, G.A. Successful Detection of the Characteristics of Tear Film Breakup Appearing Immediately after Eye Opening by Videokeratography with a Newly-Developed Indicator. *Diagnostics* **2023**, *13*, 240. [CrossRef]
44. Eftimov, P.B.; Yokoi, N.; Peev, N.; Paunski, Y.; Georgiev, G.A. Relationships between the material properties of silicone hydrogels: Desiccation, wettability and lubricity. *J. Biomater. Appl.* **2021**, *35*, 933–946. [CrossRef]
45. Pult, H.; Tosatti, S.G.; Spencer, N.D.; Asfour, J.M.; Ebenhoch, M.; Murphy, P.J. Spontaneous Blinking from a Tribological Viewpoint. *Ocul. Surf.* **2015**, *13*, 236–249. [CrossRef] [PubMed]
46. Kojima, T. Contact Lens-Associated Dry Eye Disease: Recent Advances Worldwide and in Japan. *Investig. Ophthalmol. Vis. Sci.* **2018**, *59*, DES102–DES108. [CrossRef]
47. Korb, D.R.; Greiner, J.V.; Herman, J.P.; Hebert, E.; Finnemore, V.M.; Exford, J.M.; Glonek, T.; Olson, M.C. Lid-wiper epitheliopathy and dry-eye symptoms in contact lens wearers. *Eye Contact Lens* **2002**, *28*, 211–216.
48. Shiraishi, A.; Yamaguchi, M.; Ohashi, Y. Prevalence of upper- and lower-lid-wiper epitheliopathy in contact lens wearers and non-wearers. *Eye Contact Lens* **2014**, *40*, 220–224. [CrossRef] [PubMed]
49. McMonnies, C.W. Why are soft contact lens wear discontinuation rates still too high? *Expert Rev. Ophthalmol.* **2022**, *18*, 11–18. [CrossRef]
50. Alamri, A.; Amer, K.A.; Aldosari, A.A.; Al-Muhsin, S.D.; Al-Maalwi, R.S.; Al Hamdan, S.A.; Al-Tarish, L.M. Assessment of Dry Eye Syndrome Among Contact Lens Users in Asir Region, Saudi Arabia. *Cureus* **2022**, *14*, e21526. [CrossRef]
51. Lakkis, C.; Brennan, N.A. Bulbar conjunctival fluorescein staining in hydrogel contact lens wearers. *Eye Contact Lens* **1996**, *22*, 189–194.
52. Wong, H.; Fatt, I.I.; Radke, C.J. Deposition and Thinning of the Human Tear Film. *J. Colloid Interface Sci.* **1996**, *184*, 44–51. [CrossRef]
53. Miller, K.L.; Polse, K.A.; Radke, C.J. Black-line formation and the "perched" human tear film. *Curr. Eye Res.* **2002**, *25*, 155–162. [CrossRef] [PubMed]
54. Wang, J.; Aquavella, J.; Palakuru, J.; Chung, S.; Feng, C. Relationships between central tear film thickness and tear menisci of the upper and lower eyelids. *Investig. Ophthalmol. Vis. Sci.* **2006**, *47*, 4349–4355. [CrossRef]
55. Wang, J.; Fonn, D.; Simpson, T.L.; Jones, L. Precorneal and pre- and postlens tear film thickness measured indirectly with optical coherence tomography. *Investig. Ophthalmol. Vis. Sci.* **2003**, *44*, 2524–2528. [CrossRef] [PubMed]
56. Golding, T.R.; Bruce, A.S.; Mainstone, J.C. Relationship between tear-meniscus parameters and tear-film breakup. *Cornea* **1997**, *16*, 649–661. [CrossRef]
57. Johnson, M.E.; Murphy, P.J. Temporal changes in the tear menisci following a blink. *Exp. Eye Res.* **2006**, *83*, 517–525. [CrossRef] [PubMed]
58. Guillon, M.; Dumbleton, K.; Theodoratos, P.; Patel, K.; Gupta, R.; Patel, T. Pre-contact lens and pre-corneal tear film kinetics. *Contact Lens Anterior Eye J. Br. Contact Lens Assoc.* **2019**, *42*, 246–252. [CrossRef] [PubMed]
59. Yokoi, N.; Yamada, H.; Mizukusa, Y.; Bron, A.J.; Tiffany, J.M.; Kato, T.; Kinoshita, S. Rheology of tear film lipid layer spread in normal and aqueous tear-deficient dry eyes. *Investig. Ophthalmol. Vis. Sci.* **2008**, *49*, 5319–5324. [CrossRef] [PubMed]
60. Goto, E.; Tseng, S.C. Kinetic analysis of tear interference images in aqueous tear deficiency dry eye before and after punctal occlusion. *Investig. Ophthalmol. Vis. Sci.* **2003**, *44*, 1897–1905. [CrossRef]
61. King-Smith, P.E.; Fink, B.A.; Nichols, J.J.; Nichols, K.K.; Braun, R.J.; McFadden, G.B. The contribution of lipid layer movement to tear film thinning and breakup. *Investig. Ophthalmol. Vis. Sci.* **2009**, *50*, 2747–2756. [CrossRef]
62. Owens, H.; Phillips, J. Spreading of the tears after a blink: Velocity and stabilization time in healthy eyes. *Cornea* **2001**, *20*, 484–487. [CrossRef] [PubMed]
63. Zhao, S.; Le, Q. Analysis of the first tear film break-up point in Sjögren's syndrome and non-Sjögren's syndrome dry eye patients. *BMC Ophthalmol.* **2022**, *22*, 1. [CrossRef] [PubMed]
64. Sharma, A.; Tiwari, S.; Khanna, R.; Tiffany, J.M. Hydrodynamics of meniscus-induced thinning of the tear film. *Adv. Exp. Med. Biol.* **1998**, *438*, 425–431.
65. Wong, S.; Murphy, P.J.; Jones, L. Tear evaporation rates: What does the literature tell us? *Contact Lens Anterior Eye J. Br. Contact Lens Assoc.* **2018**, *41*, 297–306. [CrossRef]
66. Werkmeister, R.M.; Alex, A.; Kaya, S.; Unterhuber, A.; Hofer, B.; Riedl, J.; Bronhagl, M.; Vietauer, M.; Schmidl, D.; Schmoll, T.; et al. Measurement of tear film thickness using ultrahigh-resolution optical coherence tomography. *Investig. Ophthalmol. Vis. Sci.* **2013**, *54*, 5578–5583. [CrossRef] [PubMed]

67. Aranha Dos Santos, V.; Schmetterer, L.; Gröschl, M.; Garhofer, G.; Schmidl, D.; Kucera, M.; Unterhuber, A.; Hermand, J.P.; Werkmeister, R.M. In vivo tear film thickness measurement and tear film dynamics visualization using spectral domain optical coherence tomography. *Opt. Express* **2015**, *23*, 21043–21063. [CrossRef]
68. Szczesna-Iskander, D.H.; Iskander, D.R. Tear film dynamics on soft contact lenses. *Optom. Vis. Sci. Off. Publ. Am. Acad. Optom.* **2014**, *91*, 1406–1411. [CrossRef] [PubMed]
69. Sharma, A.; Ruckenstein, E. Dewetting of solids by the formation of holes in macroscopic liquid films. *J. Colloid Interface Sci.* **1989**, *133*, 358–368. [CrossRef]
70. Sharma, A.; Ruckenstein, E. Energetic criteria for the breakup of liquid films on nonwetting solid surfaces. *J. Colloid Interface Sci.* **1990**, *137*, 433–445. [CrossRef]
71. Willcox, M.; Keir, N.; Maseedupally, V.; Masoudi, S.; McDermott, A.; Mobeen, R.; Purslow, C.; Santodomingo-Rubido, J.; Tavazzi, S.; Zeri, F.; et al. CLEAR—Contact lens wettability, cleaning, disinfection and interactions with tears. *Contact Lens Anterior Eye J. Br. Contact Lens Assoc.* **2021**, *44*, 157–191. [CrossRef] [PubMed]
72. Bertrand, E.; Blake, T.D.; De Coninck, J. Dynamics of dewetting. *Colloids Surf. A Physicochem. Eng. Asp.* **2010**, *369*, 141–147. [CrossRef]
73. Jones, L. Understanding the Link Between Wettability and Comfort. Contact Lens Spectr. (Improving Ocular Health and Comfort With Silicone Hydrogel Contact Lenses). 2007. Available online: https://www.clspectrum.com/supplements/2007/june-2007/improving-ocular-health-and-comfort-with-silicone/understanding-the-link-between-wettability-and-com (accessed on 1 January 2023).
74. Vidal-Rohr, M.; Wolffsohn, J.S.; Davies, L.N.; Cervino, A. Effect of contact lens surface properties on comfort, tear stability and ocular physiology. *Contact Lens Anterior Eye J. Br. Contact Lens Assoc.* **2018**, *41*, 117–121. [CrossRef]
75. Bourassa, S.; Benjamin, W.J. Clinical findings correlated with contact angles on rigid gas permeable contact lens surfaces in vivo. *J. Am. Optom. Assoc.* **1989**, *60*, 584–590. [PubMed]
76. Tonge, S.; Jones, L.; Goodall, S.; Tighe, B. The ex vivo wettability of soft contact lenses. *Curr. Eye Res.* **2001**, *23*, 51–59. [CrossRef] [PubMed]
77. Rogers, R. In Vitro and Ex Vivo Wettability of Hydrogel Contact Lenses. Master's Thesis, University of Waterloo, Waterloo, ON, Canada, 2006.

life

MDPI

Article

Effectiveness of SMILE Combined with Micro-Monovision in Presbyopic Patients: A Pilot Study

Joaquín Fernández [1], Federico Alonso-Aliste [2], Noemí Burguera [1], Julia Hernández-Lucena [2], Jonatan Amián-Cordero [2] and Manuel Rodríguez-Vallejo [1,*]

1 Qvision, Ophthalmology Department, VITHAS Almería, 04120 Almería, Spain
2 Tecnolaser Clinic Vision, Ophthalmology Department, 41018 Sevilla, Spain
* Correspondence: manuelrodriguezid@qvision.es

Abstract: Binocular summation along all defocus range after a micro-monovision procedure has scarcely been studied. The aim of this pilot study was to evaluate the efficacy of SMILE combined with different levels of micro-monovision in presbyopic patients and to assess the binocular summation effect on contrast sensitivity defocus curves (CSDC) at the 6-month follow-up. Efficacy was assessed on the basis of visual acuity (VA) and stereopsis at far, intermediate, and near distances. Patient-reported outcomes (PROs) and binocular CSDC were also evaluated. Six patients completed the study with a programmed median anisometropia of 0.81 Diopter. The median binocular uncorrected VA was better than 0 logMAR at the three evaluated distances, and stereopsis was not impaired in any patient, achieving a median of ≤119 arcsec at any distance. CSDC increased binocularly after surgery, significantly in the range of −2 to −3 D ($p < 0.05$). No clinically relevant changes were observed in PROs compared with the preoperative period, and all patients achieved spectacle independence at intermediate/near distance and were likely or very likely to undergo the same surgery. In conclusion, micro-monovision with SMILE could be an effective procedure, with results that might be comparable to other laser correction techniques specifically designed for presbyopia correction.

Keywords: presbyopia; micro-monovision; SMILE; efficacy; binocular summation

Citation: Fernández, J.; Alonso-Aliste, F.; Burguera, N.; Hernández-Lucena, J.; Amián-Cordero, J.; Rodríguez-Vallejo, M. Effectiveness of SMILE Combined with Micro-Monovision in Presbyopic Patients: A Pilot Study. *Life* **2023**, *13*, 838. https://doi.org/10.3390/life13030838

Academic Editor: Alejandro Cerviño

Received: 1 March 2023
Revised: 14 March 2023
Accepted: 17 March 2023
Published: 20 March 2023

1. Introduction

Presbyopia is a global age-related vision disorder characterized by a progressive inability to focus on nearby objects. It is estimated that about 209 million people suffer from presbyopia in Europe (44% of the population) and it is expected to affect half of the European population by 2030 [1]. Monovision is one of the available clinical techniques for correcting the visual limitations caused by presbyopia. This procedure consists of correcting the residual refractive error that causes loss of distance vision in one eye, while the other eye corrects the residual refractive defect in near vision [2]. Then, if a lower degree of anisometropia is induced (≤1.5 D), this protocol is called micro-monovision [3]. In the particular case of coexisting myopia and presbyopia, the total correction of myopia is usually applied in one eye, while the other eye is not fully corrected, leaving a residual myopia that ranges from −0.50 to −1.50 D depending on the patient's tolerance to the residual anisometropia. Complete correction of myopia in one eye and undercorrection of myopia in the other eye allows the patient to achieve better near vision than full correction of myopia in both eyes. In a monovision-adapted patient, this is because each eye dominates the other eye, according to the visual task. For example, when a patient performs tasks that require good distance vision, such as driving, the fully corrected eye dominates the undercorrected eye. On the other hand, when performing intermediate- or near-vision tasks, such as computer vision or reading, the eye undercorrected at a far distance dominates over the fully corrected eye since undercorrection favors this eye to have a higher visual acuity for the development of intermediate/near tasks. In conclusion, micro-monovision involves adapting the patient's refraction to achieve the best possible vision at all distances with the alternating use of both eyes.

Small incision lenticule extraction (SMILE) is a laser refractive surgery (LRS) technique that allows the correction of myopia and astigmatism. SMILE involves creating a lenticule inside the cornea (stroma) using a femtosecond laser, which is removed by a micro-incision of approximately 2 mm. Safety, efficacy, and predictability of SMILE for the correction of myopia and astigmatism has been widely reported, demonstrating equivalent results to laser-assisted in situ keratomileusis (LASIK), but with some advantages such as less induction of higher order aberrations and less affectation of ocular dryness [4,5].

Although micro-monovision studies have been conducted with older LRS techniques such as LASIK or photorefractive keratectomy (PRK) [2], few studies have been conducted with the combination of SMILE and micro-monovision, mainly because it is a more recent technique [6–8]. In addition, although the usual clinical practice procedure involves the correction of distance vision in the dominant eye and near vision in the non-dominant eye [6–8], recent laboratory research suggests that the vision of a patient undergoing a monovision procedure may vary according to different patterns of accommodative response [9], as presbyopic patients in their 40s and 50s still retain some accommodative capabilities. Thus, depending on the accommodative response, the patient may achieve a different range of clear vision. This pilot study aimed to evaluate the efficacy of the SMILE procedure with micro-monovision in presbyopic patients with myopia and/or astigmatism. Monocular and binocular contrast sensitivity defocus curves (CSDC) were also measured. This new metric might help to identify possible patterns that could affect the results reported by patients undergoing combined myopia, astigmatism, and presbyopia correction techniques since it measures the entire defocus range and is more sensitive than visual acuity to the decrease in optical quality due to defocus secondary to accommodation reduction [10].

2. Materials and Methods

2.1. Subjects

Consecutive patients who decided to undergo SMILE with micro-monovision for the correction of myopia and/or astigmatism and presbyopia were invited to participate in this prospective observational pilot study. The nature and possible consequences of the study were explained to all participants, who provided written informed consent. The study was approved by the Ethics Committee of Research, Almería Center, Torrecardenas Hospital Complex and adhered to the tenets of the Declaration of Helsinki.

Seven patients were recruited from October 2021 to October 2022 at two centers in Spain: Tecnolaser Clinic Vision, Sevilla (Center A, 4 patients), and Qvision, Ophthalmology Department, VITHAS Almería (Center B, 3 patients). The inclusion criteria were patients between 40 and 55 years old, presbyopia with myopia and/or myopic astigmatism ($\leq$6 D of myopia and $\leq$3 D of astigmatism), for whom micro-monovision with SMILE was programmed according to the standards of conventional clinical practice with a target in the non-dominant eye between -0.50 and -1.50 D, monocular visual acuity with best distance correction (CDVA) better or equal to 20/20, and sufficient availability, willingness, skills, and cognitive awareness to comply with follow-up/study procedures and study visits. Exclusion criteria were a history of ocular surgery (including laser refractive surgery), crystalline lens sclerosis according to the LOCS III classification system $\geq$CN1 or Pentacam Nucleus Staging (PNS) with Pentacam ≥ 2, intolerance to micro-monovision testing with contact lenses or trial frame for at least 30 min, failure of stereopsis test in near vision for all possible levels, pregnant woman at the time of surgery or follow-up, preoperative central corneal thickness less than 480 µm or a predicted postoperative residual stromal bed less than 250 µm, any ocular disorder that could potentially cause a loss of visual acuity or diplopia, topographic map compatible with subclinical keratoconus or other pathological alteration of the cornea, mu-chord >0.5 mm measured with Pentacam, use of systemic or ocular medications that may affect vision or accommodation in the last 6 months, and subjects participating in any clinical trial or research involving drugs or medical devices within 30 days prior to entry into this research and/or during the period of participation in this study.

2.2. Procedure

All patients underwent health exploration during the preoperative period for screening as candidates for micro-monovision with SMILE. This conventional exploration included slit-lamp, cycloplegic subjective refraction, corneal topography and biometry (Pentacam AXL, Oculus, Wetzlar, Germany), binocular and accommodative evaluation, photopic and mesopic pupil diameters (Keratograph 5M, Oculus, Wetzlar, Germany), and pupil diameter measured using a ruler under environmental light conditions. Ocular motor dominance was assessed by means of the pointing-a-finger test [11], then with the best distance correction in both eyes, a +1.50 D was placed over the non-dominant motor eye, reducing its value in −0.25 D steps up to achieving a tolerated value ≥ +0.50 D. This positive lens was tested for tolerance with either contact lens fitting to use at home or 30 min over the trial frame.

The study procedures included the measurement of monocular CDVA and binocular corrected visual acuities at 4 m (CDVA), 66 cm (CIVA), and 40 cm (CNVA) in the preoperative period [12], whereas monocular CDVA was measured at the 3-month safety evaluation and binocular uncorrected visual acuities at the same three distances in the 6-month follow-up for efficacy assessment (UDVA, UIVA, and UNVA). VAs were measured with an ETDRS iPad chart (VisionC, www.qvisionacademy.com, Almería, Spain) [13]. In the same way, stereopsis (StereoTAB,www.qvisionacademy.com, Almería, Spain) was evaluated at the preoperative period at far (3 m), intermediate (1.5 m), and near distances (50 cm) with best correction at each distance and in the 6-month without correction at the same distances [14]. The CSDC (MultifocalLA, www.qvisionacademy.com, Almería, Spain) was measured binocularly with the best distance correction in the preoperative period, whereas monocularly and binocularly CSDC without correction were evaluated at the 6-month follow-up. For measuring the CSDC, the patient was positioned at 4 m distance, then the MultifocalLA started with an alert message that indicated the defocus lens to be inserted (starting in +1.00 D). The experimenter pressed the orientation corresponding to the answer of the subject over a button bar of 4 possible crowded Sloan letters or a fifth button for pressing when the subject could not recognize the letter. Letters were presented randomly at a constant size of 0.3 logMAR (high spatial frequency) and increased or decreased contrast in 0.1 logCS steps depending on the patient's answer. The final threshold was automatically determined by a staircase psychophysical procedure with five reversals. After testing the threshold for the first defocus lens, a new alert appeared over the screen with the defocus lens, which replaced the previous one (from +1.00 D to +0.50 D) and the five reversals were repeated but now starting at the CS threshold obtained with the previous defocus lens. This procedure was repeated for all the defocus lenses from +1.00 D to −4.00 D, in −0.50 D steps. A complete description of this procedure can be found in the validation study in which, instead of Sloan letters, Snellen letters were used [15].

All testing measurements were taken in both centers with an iPad set up at 85 cd/m^2 of background luminance and an environmental light around 150 lux.

Patient-reported outcomes (PROs) were obtained in the preoperative stage and at 6 months by means of using the following questionnaires: the Convergence Insufficiency Symptoms Survey (CISS) to assess the near vision symptoms [16]; the Patient-Reported Spectacle Independence Questionnaire (PRSIQ) for the assessment of spectacle dependence [17]; the Vision and Night Driving Questionnaire (VND-Q) for assessing the difficulties driving at night [18]; and single questions to evaluate the dysphotopsia, and satisfaction with the procedure and desire to be submitted to the same procedure if patient had to take the decision again [19]. Adverse events were recorded during all the follow-up visits.

2.3. *Surgery*

Three experienced surgeons using the SMILE technique (JF, FAA, and JAC) performed the treatments with the VisuMax 500 kHz femtosecond laser system (Carl Zeiss Meditec AG) following the procedures of their habitual surgery practice. The final micro-monovision target in the non-dominant eye was selected depending on the tolerance test, surgeon's recommendation, and the patient's decision. Dominant eye was targeted to emmetropia and astigmatism was targeted for full correction in both eyes. The SMILE procedure involved three steps. The first step was the docking procedure, in which the center of the applanation zone was concentric with the margin of the cone and near the pupil center. In cases of astigmatism correction, a previous marking of the eye was conducted, and a slight rotation of the applanation cone was made to compensate for cyclotorsion, taking the horizontal lines seen through the microscope as a reference. After suction, the photodisruptive procedure creates a lenticule with a cap thickness between 120 and 130 µm and an optical zone diameter between 6.5 and 7.6 mm. Finally, the lenticule was extracted through an incision (2 mm wide) at the extreme end of the cap. All patients were treated preoperatively with topical anesthesia, with antibiotic and corticosteroid eye drops in the immediate postoperative period, and anti-inflammatory eye drops, decreasing the dosage progressively up to the third week after surgery.

2.4. *Statistical Analysis*

Descriptive statistics of the median (interquartile range) were selected to report the results after checking the normality using the Shapiro–Wilk test. Paired comparisons between preoperative and postoperative data were tested using the Wilcoxon signed-rank test. After closing the pilot study, a post hoc power calculation was conducted with a mean difference of 0.5 logCS and standard deviation of 0.24 logCS, obtaining a power of 0.97. SPSS software (version 26; SPSS Inc., Chicago, IL, USA) was used for the statistical analysis.

3. Results

One of the four recruited patients from center A withdrew from the study due to the inability to attend the 6-month visit. No adverse events were recorded for this patient at the last follow-up visit of 3 months and the achieved refraction was emmetropia in the dominant eye and −0.5 D in the non-dominant eye. The binocular UDVA, UIVA, and UNVA were −0.2, −0.2, and −0.1 logMAR, respectively, for this patient. Table 1 shows the demographics of the final sample of six subjects who completed all follow-up visits. The median predicted postoperative anisometropia was −0.81 D.

Table 1. Demographic characteristics of the sample.

Variable	Median	Interquartile Range
Age	45	5
Sphere (D)	−2.88	2
Cylinder (D)	−0.5	0.75
SE (D)	−3.13	1.72
Dominant Predicted SE (D)	−0.13	0.56
Non-Dominant Predicted SE (D)	−0.94	0.63
Photopic pupil diameter (mm)	3.8	1.12
Mesopic pupil diameter (mm)	6.05	0.94
Ruler pupil diameter (mm)	4	0.5
Axial Length (mm)	24.60	1.27
Mean Keratometry (D)	43.25	0.89

SE: spherical equivalent.

3.1. Safety

Only one adverse event was recorded during surgery. A small residual part of the lenticule (<1 mm) was extracted after the main lenticule was removed with no significant complications. No effects of visual acuity were observed in this eye (Table S1, Case A1). At the 3-month postoperative follow-up, a PRK procedure was conducted in the dominant eye of one patient due to undercorrection (Table S2, Case B1). The patient completed the 6-month visit and was included in the efficacy analysis. No eyes lost more than one line of CDVA at the 3-month visit compared with the preoperative period.

3.2. Efficacy

No significant differences were found for median binocular uncorrected VAs at 6 months in comparison with the best-corrected VAs at the three measured distances in the preoperative visit (Table 2). All patients maintained stereopsis without correction (Tables S1 and S2) with no significant differences (Table 2). The uncorrected binocular CSDC showed a statistically significant increase in CS from −2 to −3 D of defocus in comparison to the preoperative binocular CSDC with the best distance correction (Table 2 and Figure 1). Individual CSDC of all patients are shown in Figure S1.

Table 2. Binocular efficacy results of median (interquartile range). Preoperative with best correction at each distance and postoperative without distance correction.

Variable	Preoperative	6-Month	z, *p*-Value
Visual Acuity (logMAR)			
Far (4 m)	−0.1 (0.2)	0 (0.18)	0.7, 0.48
Intermediate (66 cm)	0 (0.02)	−0.1 (0.15)	−1.63, 0.1
Near (40 cm)	−0.1 (0.2)	−0.05 (0.18)	1.12, 0.66
Stereopsis (arcsec)			
Far (4 m)	79 (178)	119 (20)	0.32, 0.75
Intermediate (66 cm)	79 (79)	40 (9.75)	−1.63, 0.10
Near (40 cm)	79 (49)	59 (39)	−1.00, 0.32
Contrast Sensitivity (logCS)			
Defocus −4.0	0 (0.03)	0 (0.1)	1, 0.32
Defocus −3.5	0 (0.03)	0.1 (0.38)	1.63, 0.1
Defocus −3.0	0.05 (0.1)	0.5 (0.68)	2.02, 0.04
Defocus −2.5	0.15 (0.2)	0.9 (0.58)	2.04, 0.04
Defocus −2.0	0.7 (0.4)	1.2 (0.40)	2.03, 0.04
Defocus −1.5	1.05 (0.3)	1.25 (0.18)	1.60, 0.1
Defocus −1.0	1.2 (0.12)	1.15 (0.30)	0, 1
Defocus −0.5	1.25 (0.20)	1.10 (0.50)	−1.60, 0.1
Defocus 0.0	1.1 (0.38)	1.10 (0.75)	−1, 0.32
Defocus 0.5	0.55 (0.80)	0.35 (1.02)	2.22, 0.03
Defocus 1.0	0.30 (0.55)	0.05 (0.48)	−1.83, 0.07

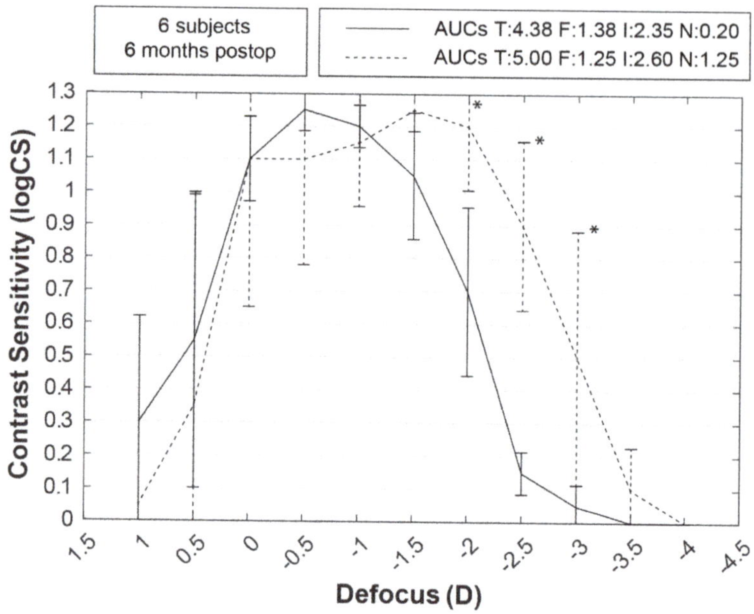

Figure 1. Preoperative best distance corrected versus postoperative uncorrected contrast sensitivity defocus curve. AUCs are the areas under the defocus curves above the 0.3 logCS value for the far (F), intermediate (I), and near (N) distance ranges. Lines represent median values, and vertical bars represent interquartile ranges. Asterisk (*) means $p < 0.05$.

3.3. Patient-Reported Outcomes

The median CISS scores were close in the preoperative, 4 (26.75), and postoperative, 3 (12), visits ($z = -1.089$, $p = 0.28$). No significant changes ($z = -0.95$, $p = 0.34$) were found for the VND-Q in the preoperative, −3.1 (4.27), and postoperative, −3.94 (6.97), periods. All patients achieved postoperative spectacle independence at intermediate and near distances, and only the patient who underwent PRK required occasional spectacle correction for far distance. Four patients were satisfied or very satisfied with their uncorrected vision at all distances, one was very satisfied with the intermediate and near vision but neutral with far vision, and one was very satisfied at intermediate but neutral at far and slightly satisfied at near. All patients were slightly or not at all bothered by photic phenomena, except for two patients who answered very bothersome and moderately bothersome, but all the patients were likely or very likely to undergo the same procedure. The individual responses of all patients are shown in Tables S1 and S2.

4. Discussion

In this pilot study, the efficacy of micro-monovision with SMILE was evaluated and binocular summation using CSDC was reported for the first time. The efficacy of micro-monovision with SMILE has been published in previous studies [6–8]. Unfortunately, none of these studies reported standardized VA measured in logMAR with an ETDRS chart; therefore, the efficacy of our study cannot be easily compared with previous micro-monovision studies with SMILE reporting results using non-standard methods for reporting VA, such as Jaeger notation or reading charts. Difficulties with making comparisons have also been found even with other micro-monovision techniques such as Presbyond, mainly for the same limitation of using Jaeger notation in these studies [20–23]. In fact, this is an important limitation of the scarce evidence of presbyopia correction with laser refractive surgery techniques [24].

Being aware of the lack of uniform testing of VA, the median UDVA and UNVA achieved in our study were 0 and −0.05 logMAR, respectively. These values were close to those reported by Reinstein et al. for UDVA (−0.07 logMAR) and UNVA (0.05 logMAR) with Presbyond and an anisometropia of 1.5 D in older patients with a median age of 55 years old [22]. On the other hand, our results were better at distance in comparison to Kohnen et al. (0.09 logMAR) and comparable for near (−0.04 logMAR) [25]. This reduction in distance vision of Kohnen's study is explained by a higher postoperative myopic spherical equivalent in the dominant eye, using the PresbyMAX with an anisometropia close to 0.5 D and older patients with a mean age of 53.4 years old [25]. Fu et al. also reported similar outcomes at near distance (0.01 logMAR) to Kohnen [26], but better distance results (−0.09 logMAR) by programming an anisometropia of 1.25 D in patients with a mean age of 47.4. According to these results, micro-monovision with SMILE appears to be a technique that might offer similar results to other platforms focused on presbyopia correction, but future randomized clinical trials are required to confirm this hypothesis.

An important finding of our study is related to the measurement of binocular uncorrected CSDC. An alternative to micro-monovision with SMILE can be the refractive lens exchange (RLE), which consists of replacing the clear crystalline lens by a multifocal intraocular lens [27]. Poorer results have been reported for RLE in young hyperopic patients (<40 years old) in comparison to our study for UDVA (0.01 logMAR), UIVA (0.2 logMAR), and UNVA (0.07 logMAR) [28]. On the other hand, binocular CSDC might result in reduced CS with multifocal IOLs [13], especially in the near and intermediate ranges, even though this has not been reported in young presbyopic patients as in our study. Considering the possible CS loss and the risks of retinal detachment in young myopic patients [27], micro-monovision of SMILE appears to be a more appropriate option, at least until the later fifties or early sixties when the retinal detachment risk decreases and the preoperative CS is lower due to crystalline lens sclerosis [29,30]. On the other hand, micro-monovision with SMILE in older patients could be questionable considering the onset of cataract development in the short term. In fact, even though the inclusion criteria was established at 55 years old in the protocol, the oldest patient in our sample was 51 years old, with a reasonable clear lens, and binocularly achieving 1 logCS in the preoperative stage at far distance, and 0.9 logCS at 0 D and −2.5 D at the 6-month visit, which means a better CS at near in comparison to that achieved with MIOLs [13]. Thus, preoperative screening considering the cut-off criteria for CS reestablishment with MIOLs is of great importance when taking the decision for laser vision correction with micro-monovision or refractive lens exchange [31]. The latter, with the consensus of the patient after explanation of the advantages and drawbacks of each procedure, will determine the best procedure for each patient. In addition, it should be noticed that patients satisfied with micro-monovision with laser vision correction could be programmed in the future for the same micro-monovision with conventional or enhanced monofocals and extended depth of focus intraocular lens, or even for multifocal intraocular lenses targeted to emmetropia [32]. Moreover, special formulas for post-laser refractive surgery such as those included in the ASCRS online calculator (https://iolcalc.ascrs.org/ accessed on 16 March 2023), thick-lens or ray-tracing formulas should be used [33,34]. Presbyopic phakic IOLs could also be an alternative for those patients, but CSDC studies are still required to determine if CS is also maintained along the whole defocus range [35–37].

An argument against monovision has been the possible loss of stereopsis [38], but previous studies with Presbyond reported minimal changes [39]. Our study is in agreement with the lack of clinically relevant differences in stereopsis loss but also provides, for the first time, evidence of intermediate and far distance stereopsis. This apparently unaltered stereopsis at all distances can be explained by programming a low anisometropia below 1.5 D [38]. Psychophysical studies have also suggested that disparity between eye images can affect reflexive eye movements, but this topic has still not been studied in clinical practice with patients operated on with laser refractive surgery and micro-monovision [40,41]. In addition, the lack of differences in symptoms evaluated with the CISS questionnaire and the VND-Q supports that micro-monovision with SMILE might not induce symptoms and difficulties in important tasks related to near work and driving at night. This, together with the low dysphotopsia incidence, explains the satisfaction rates in our study, with all patients having better satisfaction answers in comparison to the preoperative period for far and intermediate distances, and only two patients who chose one level less in the postoperative answer for near vision, with the remaining four patients selecting a better or maintained answer. Despite the reduced answers to some isolated questions, all patients agreed with the probability of being submitted again to the same procedure with a likely or very likely answer.

The main limitations of our study were its small sample size and the short-term follow-up, which limit the generalizability of the findings. However, it is important to note that the difference achieved in our study was 0.5 logCS with a micro-monovision programmed median anisometropia of 0.81 D. For this difference, and the standard deviation of 0.24 logCS obtained in this study, the power was above 0.8 for an alpha value of 0.05. Therefore, a sufficient sample size was used to avoid a Type II error at −2 D. In summary, our sample size was enough to evidence the increase in the CS at the defocus level of −2 D but was not enough to avoid a Type II error in differences below 0.4 logCS, for instance, to confirm the decrease of 0.15 logCS at −0.5 D or the increase at −1.5 D.

5. Conclusions

This was a pilot study for future estimations of the sample size based on the collected results around the main endpoint. The programmed median anisometropia of 0.8 D resulted in the effectiveness of the procedure for patients of a median age of 45 years old. The results were comparable to other presbyopia laser correction techniques that also implement micro-monovision even though this comparison should be interpreted with caution due to the differences in testing methods and age of the population. The most interesting finding was that the micro-monovision binocular summation resulted in superior contrast sensitivity as compared to the multifocal refractive lens exchange performed in a similar patient group; the stereopsis is in the normal range with the induction of this anisometropia without increasing symptoms or difficulties with near vision tasks or driving at night. On the other hand, all the findings of this pilot study should be confirmed in future studies with larger sample sizes.

Supplementary Materials: The following supporting information can be downloaded at: https://www.mdpi.com/article/10.3390/life13030838/s1, Figure S1: Contrast sensitivity defocus curves for each one of the patients at the preoperative visit with best distance correction (PreBE) and the postoperative 6-month visit without correction for right eye (PosRE), left eye (PosLE), and both eyes (PosBE); Table S1: Preoperative and postoperative results for the patients from Center A; Table S2: Preoperative and postoperative results for the patients from Center B.

Author Contributions: Conceptualization, J.F. and M.R.-V.; methodology, J.F. and M.R.-V.; formal analysis, M.R.-V.; investigation, J.F., F.A.-A., J.A.-C., N.B. and J.H.-L.; resources, J.F. and F.A.-A.; data curation, N.B., J.H.-L., J.A.-C. and M.R.-V.; writing—original draft preparation, J.F. and M.R.-V.; writing—review and editing, F.A.-A., J.A.-C., N.B. and J.H.-L.; supervision, J.F., F.A.-A. and M.R.-V.; project administration, J.F. and M.R.-V.; funding acquisition, J.F. All authors have read and agreed to the published version of the manuscript.

Funding: This research was funded by Carl Zeiss Meditec AG.

Institutional Review Board Statement: The study was conducted in accordance with the Declaration of Helsinki, and approved by the Ethics Committee of Research, Almería Center, Torrecardenas Hospital Complex (code: QVI-21-01; date: 15 September 2021).

Informed Consent Statement: Informed consent was obtained from all subjects involved in the study.

Data Availability Statement: Not applicable.

Conflicts of Interest: Indaloftal SL (Qvision) headed by Fernández received a research grant from Carl Zeiss Meditec AG for conducting this study. Carl Zeiss Meditec AG had no role in the design of the study; in the collection, analyses, or interpretation of data; in the writing of the manuscript; or in the decision to publish the results.

References

1. Marcos, S. Ocular Ageing: Improving the Quality of Sight for Cataract and Presbyopia Sufferers. *LYCHNOS (Noteb. Fund. Gen. CSIC)* **2010**, *2*, 60–65.
2. Jain, S.; Arora, I.; Azar, D.T. Success of Monovision in Presbyopes: Review of the Literature and Potential Applications to Refractive Surgery. *Surv. Ophthalmol.* **1996**, *40*, 491–499. [CrossRef] [PubMed]
3. Reinstein, D.Z.; Couch, D.G.; Archer, T.J. LASIK for Hyperopic Astigmatism and Presbyopia Using Micro-Monovision with the Carl Zeiss Meditec MEL80 Platform. *J. Refract. Surg.* **2009**, *25*, 37–58. [CrossRef] [PubMed]
4. Fu, Y.; Yin, Y.; Wu, X.; Li, Y.; Xiang, A.; Lu, Y.; Fu, Q.; Hu, T.; Du, K.; Wen, D. Clinical Outcomes after Small-Incision Lenticule Extraction versus Femtosecond Laser-Assisted LASIK for High Myopia: A Meta-Analysis. *PLoS ONE* **2021**, *16*, e0242059. [CrossRef]
5. Sambhi, R.-D.S.; Sambhi, G.D.S.; Mather, R.; Malvankar-Mehta, M.S. Dry Eye after Refractive Surgery: A Meta-Analysis. *Can. J. Ophthalmol.* **2020**, *55*, 99–106. [CrossRef]
6. Kim, J.S.; Ra, H.; Rho, C.R. Retrospective Observational Study of Micro-Monovision Small Incision Lenticule Extraction (SMILE) for the Correction of Presbyopia and Myopia. *Medicine* **2018**, *97*, e13586. [CrossRef]
7. Fu, D.; Zeng, L.; Zhao, J.; Miao, H.; Yu, Z.; Zhou, X. Safety and Satisfaction of Myopic Small-Incision Lenticule Extraction Combined with Monovision. *BMC Ophthalmol.* **2018**, *18*, 131. [CrossRef]
8. Luft, N.; Siedlecki, J.; Sekundo, W.; Wertheimer, C.; Kreutzer, T.C.; Mayer, W.J.; Priglinger, S.G.; Dirisamer, M. Small Incision Lenticule Extraction (SMILE) Monovision for Presbyopia Correction. *Eur. J. Ophthalmol.* **2017**, *28*, 287–293. [CrossRef]
9. Almutairi, M.S.; Altoaimi, B.H.; Bradley, A. Impact of Monovision on Dynamic Accommodation of Early Presbyopes. *Ophthal. Physl. Opt.* **2020**, *40*, 47–59. [CrossRef]
10. Fernández, J.; Rodríguez-Vallejo, M.; Martínez, J.; Burguera, N.; Piñero, D.P. Prediction of Visual Acuity and Contrast Sensitivity From Optical Simulations With Multifocal Intraocular Lenses. *J. Refract. Surg.* **2019**, *35*, 789–795. [CrossRef]
11. Seijas, O.; de Liaño, P.G.; de Liaño, R.G.; Roberts, C.J.; Piedrahita-Alonso, E.; Diaz, E. Ocular Dominance Diagnosis and Its Influence in Monovision. *Am. J. Ophthalmol.* **2007**, *144*, 209–216. [CrossRef] [PubMed]
12. Dupps, W.J.; Kohnen, T.; Mamalis, N.; Rosen, E.S.; Koch, D.D.; Obstbaum, S.A.; Waring, G.O.; Reinstein, D.Z.; Stulting, R.D. Standardized Graphs and Terms for Refractive Surgery Results. *J. Cataract. Refract. Surg.* **2011**, *37*, 1–3. [CrossRef]
13. Fernández, J.; Rodríguez-Vallejo, M.; Martínez, J.; Burguera, N.; Piñero, D.P. Long-Term Efficacy, Visual Performance and Patient Reported Outcomes with a Trifocal Intraocular Lens: A Six-Year Follow-Up. *J. Clin. Med.* **2021**, *10*, 2009. [CrossRef]
14. Rodríguez-Vallejo, M.; Ferrando, V.; Montagud, D.; Monsoriu, J.A.; Furlan, W.D. Stereopsis Assessment at Multiple Distances with an IPad Application. *Displays* **2017**, *50*, 35–40. [CrossRef]
15. Fernández, J.; Rodríguez-Vallejo, M.; Tauste, A.; Albarrán, C.; Basterra, I.; Piñero, D. Fast Measure of Visual Acuity and Contrast Sensitivity Defocus Curves with an IPad Application. *Open Ophthalmol. J.* **2019**, *13*, 15–22. [CrossRef]
16. González-Pérez, M.; Pérez-Garmendia, C.; Barrio, A.R.; García-Montero, M.; Antona, B. Spanish Cross-Cultural Adaptation and Rasch Analysis of the Convergence Insufficiency Symptom Survey (CISS). *Transl. Vis. Sci. Technol.* **2020**, *9*, 23. [CrossRef]
17. Morlock, R.; Wirth, R.J.; Tally, S.R.; Garufis, C.; Heichel, C.W.D. Patient-Reported Spectacle Independence Questionnaire (PRSIQ): Development and Validation. *Am. J. Ophthalmol.* **2017**, *178*, 101–114. [CrossRef]
18. Kimlin, J.A.; Black, A.A.; Wood, J.M. Older Drivers' Self-reported Vision-related Night-driving Difficulties and Night-driving Performance. *Acta Ophthalmol.* **2020**, *98*, e513–e5192020. [CrossRef]
19. Fernández, J.; Ribeiro, F.J.; Rodríguez-Vallejo, M.; Dupps, W.J.; Werner, L.; Srinivasan, S.; Kohnen, T. Standard for Collecting and Reporting Outcomes of IOL–Based Refractive Surgery: Update for Enhanced Monofocal, EDOF, and Multifocal IOLs. *J. Cataract. Refract. Surg.* **2022**, *48*, 1235–1241. [CrossRef]
20. Zhang, T.; Sun, Y.; Weng, S.; Liu, M.; Zhou, Y.; Yang, X.; Stojanovic, A.; Liu, Q. Aspheric Micro-Monovision LASIK in Correction of Presbyopia and Myopic Astigmatism: Early Clinical Outcomes in a Chinese Population. *J. Refract. Surg.* **2016**, *32*, 680–685. [CrossRef]
21. Reinstein, D.Z.; Archer, T.J.; Gobbe, M. LASIK for Myopic Astigmatism and Presbyopia Using Non-Linear Aspheric Micro-Monovision with the Carl Zeiss Meditec MEL 80 Platform. *J. Refract. Surg.* **2011**, *27*, 23–37. [CrossRef] [PubMed]
22. Reinstein, D.Z.; Carp, G.I.; Archer, T.J.; Gobbe, M. LASIK for Presbyopia Correction in Emmetropic Patients Using Aspheric Ablation Profiles and a Micro-Monovision Protocol With the Carl Zeiss Meditec MEL 80 and VisuMax. *J. Refract. Surg.* **2012**, *28*, 531–541. [CrossRef] [PubMed]
23. Ganesh, S.; Brar, S.; Gautam, M.; Sriprakash, K. Visual and Refractive Outcomes Following Laser Blended Vision Using Non-Linear Aspheric Micro-Monovision. *J. Refract. Surg.* **2020**, *36*, 300–307. [CrossRef]
24. Fernández, J.; Molina-Martín, A.; Rocha-de-Lossada, C.; Rodríguez-Vallejo, M.; Piñero, D.P. Clinical Outcomes of Presbyopia Correction with the Latest Techniques of PresbyLASIK: A Systematic Review. *Eye* **2022**, *37*, 1–10. [CrossRef] [PubMed]

25. Kohnen, T.; Böhm, M.; Herzog, M.; Hemkeppler, E.; Petermann, K.; Lwowski, C. Near Visual Acuity and Patient-Reported Outcomes in Presbyopic Patients after Bilateral Multifocal Aspheric Laser in Situ Keratomileusis Excimer Laser Surgery. *J. Cataract. Refr. Surg.* **2020**, *46*, 944–952. [CrossRef]
26. Fu, D.; Zhao, J.; Zeng, L.; Zhou, X. One Year Outcome and Satisfaction of Presbyopia Correction Using the PresbyMAX®Monocular Ablation Profile. *Front. Med.* **2020**, *7*, 589275. [CrossRef]
27. Nanavaty, M.A.; Daya, S.M. Refractive Lens Exchange versus Phakic Intraocular Lenses. *Curr. Opin. Ophthalmol.* **2012**, *23*, 54–61. [CrossRef]
28. Djodeyre, M.R.; Ortega-Usobiaga, J.; Beltran, J.; Druchkiv, V.; Baviera-Sabater, J.; Bouza-Miguens, C. Bilateral Refractive Lens Exchange With Trifocal Intraocular Lens for Hyperopia in Patients Younger Than 40 Years: A Case–Control Study. *J. Refract. Surg.* **2021**, *37*, 524–531. [CrossRef]
29. Thylefors, J.; Jakobsson, G.; Zetterberg, M.; Sheikh, R. Retinal Detachment after Cataract Surgery: A Population-based Study. *Acta Ophthalmol.* **2022**, *100*, e1595–e15992022. [CrossRef]
30. Fernández, J.; Rodríguez-Vallejo, M.; Martínez, J.; Tauste, A.; Piñero, D.P. From Presbyopia to Cataracts: A Critical Review on Dysfunctional Lens Syndrome. *J. Ophthalmol.* **2018**, *2018*, 4318405. [CrossRef]
31. Fernández, J.; Burguera, N.; Rocha-de-Lossada, C.; Rachwani-Anil, R.; Rodríguez-Vallejo, M. Objective Cataract Grading Methods and Expected Contrast Sensitivity Reestablishment with Multifocal Intraocular Lenses. *Int. Ophthalmol.* **2023**, 1–8. [CrossRef]
32. Sevik, M.O.; Turhan, S.A.; Toker, E. Clinical Outcomes with a Low Add Multifocal and an Extended Depth of Focus Intraocular Lenses Both Implanted with Mini-Monovision. *Eye* **2021**, *36*, 1168–1177. [CrossRef]
33. Fernández, J.; Rodríguez-Vallejo, M.; Martínez, J.; Tauste, A.; Piñero, D.P. New Approach for the Calculation of the Intraocular Lens Power Based on the Fictitious Corneal Refractive Index Estimation. *J. Ophthalmol.* **2019**, *2019*, 2796126. [CrossRef]
34. Lwowski, C.; Keer, K.V.; Adas, M.; Schwarz, L.; Hinzelmann, L.; Pawlowicz, K.; Kohnen, T. Ray-Tracing Calculation Using Scheimpflug Tomography of Diffractive Extended Depth of Focus IOLs Following Myopic LASIK. *J. Refract. Surg.* **2021**, *37*, 231–239. [CrossRef]
35. Schmid, R.; Luedtke, H. A Novel Concept of Correcting Presbyopia: First Clinical Results with a Phakic Diffractive Intraocular Lens. *Clin. Ophthalmol. Auckl. N. Z.* **2020**, *14*, 2011–2019. [CrossRef] [PubMed]
36. BV, O. Clinical Trial With Artiflex Presbyopic. Available online: https://clinicaltrials.gov/ct2/show/NCT04632784 (accessed on 14 March 2023).
37. Packer, M.; Alfonso, J.F.; Aramberri, J.; Elies, D.; Fernandez, J.; Mertens, E. Performance and Safety of the Extended Depth of Focus Implantable Collamer®Lens (EDOF ICL) in Phakic Subjects with Presbyopia. *Clin. Ophthalmol. Auckl. N. Z.* **2020**, *14*, 2717–2730. [CrossRef] [PubMed]
38. Fawcett, S.L.; Herman, W.K.; Alfieri, C.D.; Castleberry, K.A.; Parks, M.M.; Birch, E.E. Stereoacuity and Foveal Fusion in Adults with Long-Standing Surgical Monovision. *J. Am. Assoc. Pediatric. Ophthalmol. Strabismus* **2001**, *5*, 342–347. [CrossRef]
39. Romero, M.; Castillo, A.; Carmona, D.; Palomino, C. Visual Quality after Presbyopia Correction with Excimer Laser Ablation Using Micromonovision and Modulation of Spherical Aberration. *J. Cataract. Refract. Surg.* **2019**, *45*, 457–464. [CrossRef] [PubMed]
40. Quaia, C.; Optican, L.M.; Cumming, B.G. Binocular Summation for Reflexive Eye Movements. *J. Vis.* **2018**, *18*, 7. [CrossRef]
41. Boström, K.J.; Warzecha, A.-K. Ocular Following Response to Sampled Motion. *Vis. Res.* **2009**, *49*, 1693–1701. [CrossRef]

life

MDPI

Article

Key Factors in Early Diagnosis of Myopia Progression within Ocular Biometric Parameters by Scheimpflug Technology

Alfredo López-Muñoz [1,2], Beatriz Gargallo-Martínez [1], María Carmen Sánchez-González [1], Raúl Capote-Puente [1], Concepción De-Hita-Cantalejo [1], Marta Romero-Luna [1], Juan-José Conejero-Domínguez [1] and José-María Sánchez-González [1,*]

[1] Department of Physics of Condensed Matter, Optics Area, Vision Sciences Research Group (CIVIUS), Pharmacy School, University of Seville, 41009 Seville, Spain

[2] Research & Development Department (Miranza Virgen de Luján®), Ophthalmology Center, 41011 Seville, Spain

* Correspondence: jsanchez80@us.es; Tel.: +34-955420861

Abstract: The aim of this study was to evaluate the relationship between myopia and ocular biometric variables using the Pentacam AXL® single rotation Scheimpflug camera. This prospective, cross-sectional, single-center study was performed in fifty Caucasian patients aged between 18 and 30 years (24.84 ± 3.04 years). The measured variables included maximum and minimum keratometry (K1 and K2, respectively), anterior chamber depth (ACD), corneal horizontal diameter or white to white (WTW), central corneal thickness (CCT), corneal asphericity (Q), and axial length (AXL). The tomographic and biometric measurements were considered optimal when the quality factor was greater than 95% according to the manufacturer's software instructions. The AXL presented a significant correlation with the spherical equivalent without cycloplegia (SE without CP), age at onset of myopia ($r = -0.365$, $p = 0.012$), mean keratometry (Km) ($r = -0.339$, $p = 0.016$), ACD ($r = 0.304$, $p = 0.032$), and WTW ($r = 0.406$, $p = 0.005$). The eyes with AXL higher than 25 mm had earlier onset; higher SE without CP, AXL, and Q; and a flatter Km. AXL is the biometric variable with the greatest influence on the final refractive state in the adult myopic eye. Ophthalmologists and optometric management must consider these biometric differences in order to identify the most appropriate correction techniques in each case. The use of the Pentacam AXL in ocular biometric measurement is effective, reproducible, and non-invasive.

Keywords: axial length; ocular biometry; early diagnosis; myopia progression; Scheimpflug technology

Citation: López-Muñoz, A.; Gargallo-Martínez, B.; Sánchez-González, M.C.; Capote-Puente, R.; De-Hita-Cantalejo, C.; Romero-Luna, M.; Conejero-Domínguez, J.-J.; Sánchez-González, J.-M. Key Factors in Early Diagnosis of Myopia Progression within Ocular Biometric Parameters by Scheimpflug Technology. *Life* **2023**, *13*, 447. https://doi.org/10.3390/life13020447

Academic Editor: Wallace B. Thoreson

Received: 4 January 2023
Revised: 2 February 2023
Accepted: 3 February 2023
Published: 4 February 2023

1. Introduction

Myopia is the most common clinically significant refractive defect and is responsible for 5–10% of cases of legal blindness in developed countries [1]. Myopia, also known as nearsightedness, is a prevalent public health concern that can lead to visual impairment and increase the risk of other serious eye conditions. It is widely recognized as a significant issue [2]. Epidemiological studies conducted in recent years have detected a rising prevalence of myopia globally. Some projections estimate myopia will affect 34% of the world's population in 2020, with an asymmetric distribution between different geographical regions and ethnic groups. In Western Europe, according to some publications, myopia affects approximately 30–35% of people in 2020 [1]. Myopia is often first diagnosed in childhood and can progress throughout adolescence. It is more common in people of East Asian descent and tends to run in families. In some countries in these regions, over 80% of the population has myopia. The prevalence of myopia is also increasing in other parts of the world, including the United States and Europe [3].

There are several risk factors for the development of myopia, including genetics, increased near work (e.g., reading, using a computer), less time spent outdoors, and higher levels of education. Myopia is associated with a number of health complications, including an increased risk of retinal detachment, glaucoma, and cataracts. It can also lead to visual impairment and reduced quality of life [4]. In Spain, the occurrence of myopia appears to be on the rise. It is believed that lifestyle factors may be contributing to the increased likelihood of developing myopia [5]. Several studies suggest the importance of understanding the mechanisms responsible for the progression of myopia, and effective treatments that may slow or prevent the onset of myopia in children are being researched [6].

Axial length, or the length of the eye, is considered to be a significant factor in the development of myopia. When the length of the vitreous chamber (a fluid-filled space in the eye) exceeds the focal length of the eye's optical components, it can increase the refractive power of the eye and contribute to the development of refractive errors, such as myopia [7]. Genetic factors play a significant role in the development and progression of myopia, which are also directly related to environmental factors of all activities carried out requiring near vision [8]. There are four structures in the eye that contribute to its refractive status: the cornea, aqueous humor, lens, and vitreous humor. When these structures do not work together properly, it can lead to refractive errors such as myopia. Factors such as corneal curvature, anterior chamber depth, lens thickness, vitreous chamber depth, and axial length are often studied in relation to eye diseases, as they can impact the coordination of the ocular components and contribute to the development of refractive disorders [4].

There have been recent advances in the development of devices for the measurement of ocular biometrics, which are essential for studying and controlling myopia [9,10]. These devices allow for the precise and non-invasive measurement of various ocular parameters, such as axial length, corneal curvature, and lens thickness. There is ongoing development of devices to more accurately measure ocular biometrics such as corneal curvature and lens thickness. Currently, several non-invasive devices are used in clinical settings to measure these parameters, including partial coherence interferometry, optical low coherence reflectometry, and swept source biometry. These devices allow for the analysis of relationships between different ocular components and facilitate the use of different measurement systems [11].

When talking about a clinical diagnostic test, parameters such as sensitivity, specificity, and positive and negative predictive values are described. These reflect the characteristics of a diagnostic test and are used to decide when they should be used (sensitivity and specificity of a test) or what meaning a test result has in a particular patient.

Sensitivity is the probability of correctly classifying patients or, what is the same, the proportion of true positives, while the specificity is the probability of correctly classifying the healthy ones or, what is the same, the proportion of true negatives. Accordingly, the sensitivity and specificity represent the validity of a diagnostic test and the positive predictive value and negative predictive value represent the safety of a diagnostic test [12,13].

The Pentacam AXL® device, made by Oculus in Germany, combines the Scheimpflug principle with partial coherence interferometry to measure ocular biometrics, including anterior segment tomography, axial length, and intraocular lens calculations and it is an indispensable tool, with high values of sensitivity and specificity in diagnosis. It uses a blue LED light source with a wavelength of 475 nm and a 1.45-megapixel camera to capture images of the cornea and record 138,000 datapoints in 2 s. Keratometry is calculated using a reference surface. This device builds on the proven measurement method of the Pentacam HR [14]. Other new devices for the measurement of ocular biometrics include swept source biometry and optical low coherence reflectometry. These devices use different technologies to measure ocular parameters, but they have also been found to be accurate and reliable. The use of these new devices for the measurement of ocular biometrics has allowed for a greater understanding of the factors that contribute to the development and progression of myopia. This knowledge can be used to develop effective interventions and treatments for myopia control [11].

The purpose of this study was to examine the connection between myopia and ocular biometric measurements taken using the Pentacam AXL® device, a single rotation Scheimpflug camera with version 6.08r19 software produced by Oculus Optikgeräte in Germany.

2. Materials and Methods

2.1. Design

This research was conducted as a prospective, cross-sectional, single-center study at the University of Seville's School of Pharmacy in Spain from February to April 2019. It followed the guidelines of the Declaration of Helsinki and was approved by the Ethical Committee Board of Andalusia. All study participants provided informed consent after being informed about the nature of the study.

2.2. Subjects

Fifty Caucasian patients aged 18 to 30 years (mean age: 24.84 ± 3.04 years) who were students at the University of Seville were recruited for the study [15]. To eliminate bias, only one eye of each participant was randomly chosen for inclusion. The inclusion criteria were as follows: being between 18 and 30 years old, having a stable refractive error (no more than 0.50 D change in spherical and cylindrical refraction in the past year), having a simple or compound myopic refractive error with or without astigmatism, having a corrected visual acuity of at least 20/25 in both eyes, and not wearing contact lenses for at least 2 weeks (4 weeks for hard lenses). The exclusion criteria were as follows: having any eye disease (e.g., glaucoma, cataracts), progressive corneal disease (e.g., keratoconus, pellucid marginal degeneration), corneal dystrophy or degeneration, cataracts or sclerosis of the lens, a current or previous history of uveitis, dry eye syndrome, persistent epithelial defects, central corneal leucoma, being on antiglaucoma or hypotonic therapy, signs of retinal vascular pathology, being pregnant or lactating, and having disorders of the eye muscles (e.g., strabismus, nystagmus) or any other disorder that affects ocular fixation.

2.3. Procedure

The participants underwent cycloplegic autorefraction and traditional refraction using retinoscopy and then subjective refraction with fogging was performed. Fogging refers to using plus powers to bring the optical point of focus in front of the retina to ensure that accommodation is adequately relaxed. The principle of fogging involves using spherical powers to create artificial myopia, thereby moving the entire area of focus in the eye in front of the retina to create a situation where an attempt at accommodating will blur the vision, which further causes the patient to relax accommodation. Fogging is effective irrespective of the inherent refractive state of the eye and the efficacy of fogging in refraction has been demonstrated [16].

Corneal tomography, analysis of the anterior segment, and measurement of axial length using the Pentacam AXL® device were then carried out. The variables studied were maximum and minimum keratometry, anterior chamber depth (ACD), corneal horizontal diameter (WTW), central corneal thickness, corneal asphericity (Q), and axial length (AXL). The mean keratometry (Km) was calculated as the average of K1 and K2 within the 3-mm central optical zone for statistical analysis. A flowchart of the design of the study is presented in Figure 1.

Each measurement was taken three times in succession by the same experienced examiner under standardized conditions to minimize error. The tomographic and biometric measurements were considered optimal when the quality factor was greater than 95% according to the manufacturer's software instructions.

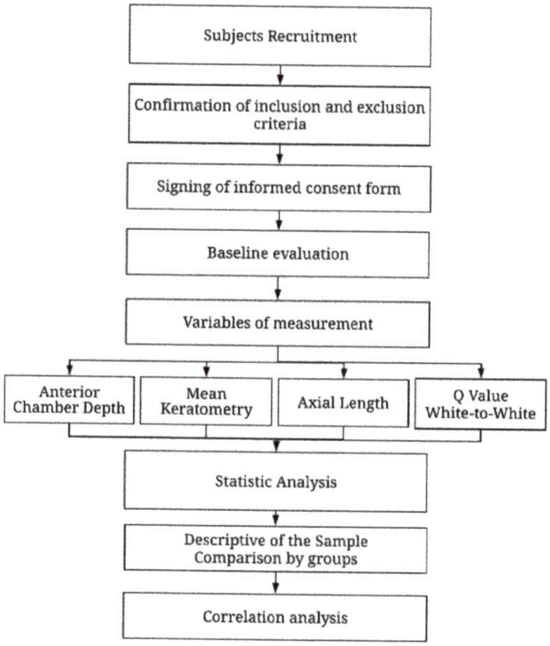

Figure 1. Research procedure flowchart.

2.4. Data Analysis

Statistical analysis was performed using SPSS software, version 23.0 (SPSS Inc., IBM, Chicago, IL, USA). Data distribution normality was studied by the Shapiro–Wilk test. Parametric variables (age, onset of myopia, AXL, Q, pachymetry, and ACD) and non-parametric variables (SE, Km, and WTW) were correlated using the Pearson coefficient correlation test and rho Spearman test, respectively. The sample was divided into two groups based on the dependent variables. The cutoff values (SE without CP, −4.50 D; AXL, 25 mm; age at myopia onset, 11 years) were arbitrarily set based on the median value. The mean difference in variables between these groups was analyzed using the independent samples Student's *t*-test (parametric), Mann–Whitney U test (non-parametric), or Chi-Square test (categorical variables). Finally, a multiple stepwise linear regression was performed. $p \leq 0.05$ was considered statistically significant.

3. Results

Fifty eyes from 50 patients (mean age, 24.84 ± 3.04 years, range 18–30 years) were included. There were 21 men (42%) and 29 women (58%) and no differences were found between sex groups for any variable analyzed. The descriptive data for the whole sample are presented in Table 1. The SE without CP showed a strong significant correlation with AXL (Figure 2) and age at myopia onset (Figure 3). Eyes with myopic SE without CP higher than −4.5 D had significantly longer AXL and younger age at myopia onset (Table 2).

No other ocular variables showed significant differences between SE without CP groups. The AXL was significantly correlated with the SE without CP (Figure 1), age at myopia onset (r = −0.365, p = 0.012), Km (r = −0.339, p = 0.016), ACD (r = 0.304, p = 0.032), and WTW (r = 0.406, p = 0.005). The eyes with AXL greater than 25 mm showed an earlier onset; higher values of SE without CP, AXL, and Q; and a flatter Km (Table 2). Age at myopia onset was significantly correlated with SE without CP (Figure 2), AXL (r = −0.365, p = 0.012), and Q (r = 0.371, p = 0.010). Patients with onset of myopia earlier than 11 years had higher SE without CP, AXL, and Q values (Table 2).

Table 1. Descriptive analysis of the sample.

Variables			
	Mean ± SD	95% CI Lower	95% CI Higher
Age (years)	24.84 ± 3.04	24.42	26.12
Onset of Myopia (years)	10.83 ± 4.48	9.68	12.37
SE without CP (D)	−5.16 ± 2.72	−6.05	−4.36
AXL (mm)	25.40 ± 1.20	24.97	25.71
Km (D)	43.87 ± 1.25	43.50	44.28
Q	−0.34 ± 0.10	−0.37	−0.34
Pachymetry (µm)	525.09 ± 30.42	515.95	534.23
ACD (mm)	3.32 ± 0.27	3.23	3.40
WTW (mm)	11.96 ± 0.34	11.85	12.05

SE = Sphere Equivalent; CP = Cycloplegia; AXL = Axial Length; Km = Mean anterior corneal keratometry; Q = Asphericity; ACD = Anterior Chamber Depth; WTW = Horizontal corneal diameter; D = Diopters; mm = millimeters; µm = micrometers; SD = Standard Deviation.

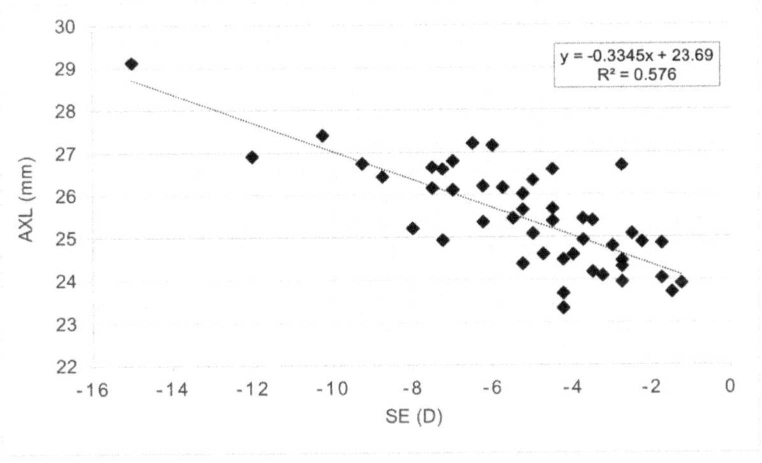

Figure 2. Correlation graph between spherical equivalent and AXL. Axial length (AXL), Spherical equivalent (SE) without cycloplegia and diopters (D).

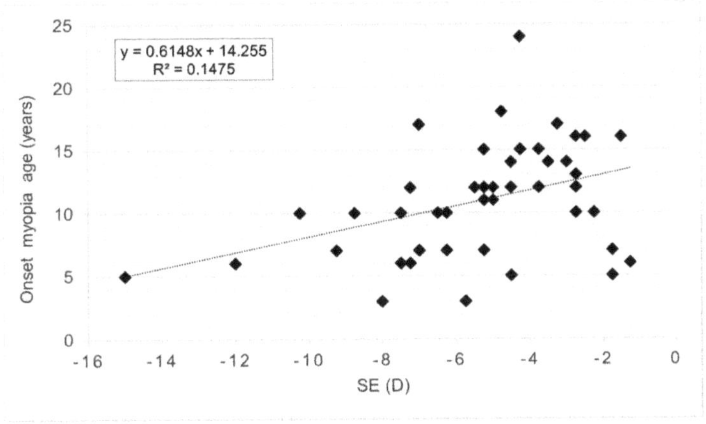

Figure 3. Correlation graph between onset of myopia age and spherical equivalent. Spherical equivalent (SE) without cycloplegia (CP) and diopters (D).

Table 2. Differences in the demographic and ocular parameters between groups by the SE, AXL, and onset of myopia age.

	SE			AXL			Onset Myopia Age			Statistical Test
	≤−4.5D	>−4.5D	p	≤25 mm	>25 mm	p	≤11 years	>11 years	p	
n	28	22		22	28		25	25		
Males/Females (n)	11/17	10/12	0.440	8/14	13/15	0.335	9/16	12/13	0.284	1
Age (years)	24.57 ± 3.27	25.18 ± 2.77	0.488	25.45 ± 2.72	24.36 ± 3.25	0.209	24.92 ± 3.46	24.76 ± 2.63	0.855	2
Onset of Myopia (years)	9.22 ± 3.52	13.00 ± 4.79	0.003 *	9.61 ± 3.31	12.33 ± 5.30	0.037 *	7.44 ± 2.42	14.68 ± 2.83	<0.001 *	2
SE without CP (D)	−6.87 ± 2.43	−2.99 ± 0.90	<0.001 *	−3.44 ± 1.43	−6.51 ± 2.74	<0.001 *	−6.60 ± 2.89	−3.72 ± 1.54	<0.001 *	3
AXL (mm)	26.09 ± 1.04	24.52 ± 0.73	<0.001 *	24.32 ± 0.44	26.25 ± 0.89	<0.001 *	25.96 ± 1.28	24.85 ± 0.82	0.001	2
Km (D)	43.82 ± 1.22	43.80 ± 1.31	0.822	44.43 ± 1.44	43.43 ± 0.87	0.016 *	43.97 ± 1.47	43.77 ± 1.01	0.459	3
Q	−0.35 ± 0.10	−0.32 ± 0.10	0.230	−0.30 ± 0.09	−0.37 ± 0.10	0.023 *	−0.37 ± 0.11	−0.31 ± 0.09	0.047	2
Pachymetry (µm)	526,82 ± 29.67	525.95 ± 29.55	0.919	522.73 ± 29.79	529.36 ± 29.14	0.433	526.36 ± 29.37	526.52 ± 29.86	0.985	2
ACD (mm)	3.35 ± 0.26	3.28 ± 0.28	0.289	3.27 ± 0.25	3.37 ± 0.28	0.214	3.34 ± 0.25	3.31 ± 0.29	0.748	2
WTW (mm)	11.99 ± 0.30	11.93 ± 0.40	0.582	11.83 ± 0.42	12.07 ± 0.22	0.030 *	11.91 ± 0.31	12.01 ± 0.37	0.207	3

SE = Sphere Equivalent; CP = Cycloplegia; AXL = Axial Length; Km = Mean anterior corneal keratometry; Q = Asphericity; ACD = Anterior Chamber Depth; WTW= Horizontal corneal diameter; D = Diopters; mm = millimeters; µm = micrometers. Statistical test performed: Chi-square test (1), Independent sample Student's t-test (2) and Mann–Whitney U test (3).
* Statistically significant data ($p < 0.05$).

The multiple linear regression results showed that the variables most closely related to SE were AXL ($p < 0.001$, partial regression coefficient B = −2.441; standardized coefficient beta = −1.066), Km ($p < 0.001$, partial regression coefficient B = −1.021, standardized coefficient beta = −0.472), and ACD ($p < 0.001$, partial regression coefficient B = 3.616, standardized coefficient beta = 0.353); adjusted $R^2 = 0.917$. A second multiple linear regression for each SE without CP group was performed, and different results were found. In the lower SE without CP group (SE > −4.5 D), a limited model (adjusted $R^2 = 0.474$) showed the most closely related variable to SE without CP was the age at myopia onset ($p = 0.011$, standardized coefficient beta = 0.502), followed by the Km ($p = 0.015$, standardized coefficient beta = −0.478). In these cases, the AXL score was not significantly related ($p = 0.232$).

However, in the higher myopic SE without CP group (SE ≤ −4.5 D), the best model (adjusted $R^2 = 0.877$) showed that the most closely related variables to SE without CP were AXL ($p < 0.001$, standardized coefficient beta = −1.102), Km ($p < 0.001$, standardized coefficient beta = −0.530), and ACD ($p < 0.001$, standardized coefficient beta = 0.455).

4. Discussion

The purpose of this study was to evaluate the relationship between myopia and ocular biometric variables using the Pentacam AXL single rotation Scheimpflug camera. This non-invasive measurement instrument was used to assess a range of variables, including maximum and minimum keratometry, anterior chamber depth, corneal horizontal diameter, central corneal thickness, corneal asphericity, and axial length. The results of the study showed that axial length (AXL) was significantly correlated with a number of other variables, including the spherical equivalent, age at onset of myopia, mean keratometry, anterior chamber depth, and corneal horizontal diameter. The eyes with AXL values higher than 25 mm were found to have earlier onset of myopia, higher spherical equivalent, AXL, and corneal asphericity values, and flatter mean keratometry values. These findings suggest that AXL is a key biometric variable that should be considered in the management of myopia, as it has a significant influence on the final refractive state of the adult myopic eye.

In recent years, there has been great interest in precisely knowing the dimensions of the human eye. Several studies of ocular biometrics have been conducted, but most have been performed in the elderly [10,17] or in subjects with a wide range of ages [18,19]. To minimize the impact of aging on ocular biometric measurements, it is necessary to carefully design studies that take this factor into account [20]. There have been numerous studies that have compared the values of variables in the anterior segment of myopic and emmetropic eyes. These studies have consistently found significant correlations between refractive errors and anterior chamber depth [17], corneal diameter [18,21], Km [22], and PQT [23]. The strength of correlations between refractive errors and anterior chamber depth may vary depending on the race and age of the study subjects. It is also known that the prevalence of myopia can differ based on age, ethnicity, and educational level, with higher rates being reported in university students in Asian countries [24]. To the best of our knowledge, there are no published studies on the correlations between the degree of myopia and the biometrics of the anterior segment in young Spanish university students.

Most studies that analyze university populations focus on the analysis of the prevalence of refractive errors by measuring only refraction [25], or only measuring AXL [26], while others use two devices for different measurements in the analysis of different ametropias [20]. One of the advantages of our study is the use of a single, non-invasive device, with which the different variables are obtained in a single measurement, such as the Pentacam AXL®, whose values of AXL, Km, and ACD have been shown to be highly repeatable [27].

There are various tools available for the measurement of axial length, including ultrasound [28], partial coherence laser interferometry [29], and optical coherence tomography [30]. Each of these tools has its own unique advantages and disadvantages. Ultrasound is a non-invasive method of measuring axial length, but its accuracy can be affected by factors such as the thickness of the cornea, presence of cataracts, and the presence of media opacities [28]. On the other hand, partial coherence laser interferometry provides highly accurate measurements, but requires a dilated pupil and can be affected by corneal distortion [29].

Optical coherence tomography (OCT) is a non-contact, non-invasive technique that provides high-resolution cross-sectional images of the retina and anterior segment. It is widely considered to be the most accurate method for measuring axial length, but it can be affected by small amounts of corneal reflection and scatter [30]. The Pentacam measures axial length using optical low coherence reflectometry (OLCR) technology. This is a non-contact, non-invasive method of measuring the axial length of the eye. The measurement is performed by shining a light into the eye and analyzing the reflections that are returned. The measurement is typically performed in a matter of seconds, providing a quick and easy method of measuring axial length. The Pentacam has become a commonly used tool in ophthalmology for the assessment of eye anatomy and for guiding surgical procedures such as cataract and refractive surgery [31]. The choice of tool for measuring the axial length of the eye depends on various factors, including the accuracy desired, the presence of any ocular or systemic factors that may affect the measurement, and the need for non-invasive or non-contact methods. It is important to consider the advantages and disadvantages of each tool when choosing the most appropriate method for a specific patient or research study.

The Pentacam offers several advantages over traditional methods for calculating the axial length (AXL) of the eye. Some of these advantages include: non-contact technology [32]: the Pentacam uses a rotating Scheimpflug camera to capture images of the eye, which eliminates the need for contact with the eye and reduces the risk of corneal damage or infection [28]. Second, high precision: the Pentacam uses laser-based technology to accurately measure the AXL, providing highly precise and reliable results [33]. Third, comprehensive analysis: the Pentacam not only measures the AXL but also provides a comprehensive analysis of the anterior and posterior segments of the eye, allowing for a more complete understanding of the ocular structure and function. Finally, automated process: the Pentacam has an automated process that eliminates the need for manual measurements, reducing the risk of measurement errors and providing more accurate results [34].

Our results showed that axial length, mean keratometry, and anterior chamber depth had significant correlations with myopic refractive error, particularly in subjects with axial lengths greater than 25 mm. A longer axial length is typically associated with a greater degree of myopia, but a flatter cornea and deeper anterior chamber depth can reduce the refractive error. A less flat cornea leads to lower refractive power, resulting in the light focusing on a point further from the cornea (i.e., hyperope displacement). When the anterior chamber depth increases, the distance between the cornea and lens increases, which also leads to a hyperopic change. These compensatory changes can result in emmetropization. Our results are consistent with other studies that have found flatter corneas in myopia [24,35].

It is currently unknown if the biological processes underlying the development of myopia at the age of 7 differ from those that occur during the early adult years, so the age of onset of myopia may not be a reliable indicator [2]. It has also been shown that structural complications in individuals with severe myopia are highly dependent on age [36]. Chua et al. found that the age at the onset of myopia, or the duration of its progression, was the most important predictor in cases of severe myopia in myopic children [3]. Our results agree with these findings, since patients who had developed myopia before age 11 had higher values of SE without CP, AXL, and Q. Regarding Q, our study is consistent with that of Horner et al., who suggested that young nearsighted individuals with more prolate corneas tended to develop higher degrees of juvenile myopia [37], as opposed to the study performed by Yebra-Pimentel et al., who found no correlation between Q and refractive error [38]. Although previous studies have reported a significant correlation between PQT and refractive error [39], our results suggest that the correlation between PQT and myopia is very weak. Several population studies have reported that men were slightly but significantly more nearsighted than women [40,41]. In our study, there were no significant differences between men and women for any of the variables.

The strengths of this study include the use of the Pentacam AXL® device, a reliable and accurate tool for measuring ocular biometrics, and the prospective, cross-sectional design, which allowed for the evaluation of the relationship between myopia and ocular biometric variables in a large sample of patients. The study also included a wide range of variables, including maximum and minimum keratometry, anterior chamber depth, corneal horizontal diameter, central corneal thickness, corneal asphericity, and axial length. This design offers several advantages, including time-efficiency, as data are collected at one point in time; low cost, as no follow-up is needed; and suitability for large sample sizes, as the design can accommodate a large sample size. Compared to other methodologies, this design is different as it is prospective, collecting data in advance, and cross-sectional, collecting data at one point in time, instead of retrospectively or repeatedly over time. Additionally, this study was conducted at a single location, unlike multi-center designs which are conducted at multiple locations. The measurements were taken multiple times to minimize error and the quality factor of the tomographic and biometric measurements was required to be greater than 95% to be considered optimal. The results of the study showed significant correlations between axial length and several other variables, including spherical equivalent, age at onset of myopia, mean keratometry, anterior chamber depth, and corneal horizontal diameter. This suggests that axial length is an important factor in the development and progression of myopia and that it may be useful in identifying the most appropriate correction techniques for patients. Overall, the study provides valuable insights into the relationship between myopia and ocular biometric variables, and demonstrates the effectiveness of the Pentacam AXL® in measuring these variables. These findings have important implications for the management of myopia and may inform the development of strategies for its control.

There are several limitations to this study that should be considered when interpreting the results. First, the sample size is relatively small, with only 50 patients included in the study. This may not be representative of the larger population and could affect the generalizability of the findings. Second, the study is cross-sectional in nature, which means that it only captures a snapshot of the data at a single point in time. This precludes the ability to establish any causal relationships between the variables of interest. Third, the study is limited to a single center, which may not be representative of the diversity of patient populations seen in other settings. Finally, the study is limited to a specific age range of 18–30 years, which means that the results may not be applicable to individuals outside of this age range. In addition, all patients reported approximately the age of onset of myopia, but their medical history was not consulted the first time they underwent refraction.

In addition to the limitations already mentioned, there are a few other factors to consider when interpreting the results of this study. First, the sample is restricted to Caucasian patients, which means that the results may not be generalizable to other racial or ethnic groups. Second, the study only included patients with myopia, which means that the results may not be applicable to individuals with other types of refractive errors. Third, the measurement of ocular biometric variables was performed using the Pentacam AXL single rotation Scheimpflug camera, which is a specific type of instrument with its own set of limitations and potential sources of error. Fourth, the cycloplegic refraction was not performed. Cycloplegia is important because it temporarily paralyzes the accommodation reflex of the eye, this is particularly useful in determining the appropriate prescription for corrective lenses [42]. When the spherical equivalent is calculated, several factors should be considered, including the patient's age, their level of accommodative ability, and any previous prescription for corrective lenses [42]. Finally, the study did not examine the influence of environmental or genetic factors on the relationship between myopia and ocular biometric variables, which could be important considerations in future research.

There are several directions for future research that could build upon the findings of this study. First, it would be interesting to replicate this study in a larger and more diverse sample to confirm the results and determine the generalizability of the findings to other populations. Second, a longitudinal study design, in which the same patients are followed over time, could provide insights into the temporal relationships between myopia and ocular biometric variables. Third, the influence of environmental and genetic factors on the relationship between myopia and ocular biometric variables could be examined in future research. This could be achieved by collecting data on patient exposures and family history and using statistical techniques to control for these factors. Finally, the use of other measurement instruments or techniques, such as ultrasound or magnetic resonance imaging, could provide additional insights into the relationship between myopia and ocular biometric variables.

5. Conclusions

AXL is the biometric variable with the greatest influence on the final refractive state in the adult myopic eye. Ophthalmologists and optometric management must consider these biometric differences in order to identify the most appropriate correction techniques in each case. The use of the Pentacam AXL in ocular biometric measurement is effective, reproducible, and non-invasive. The results of our study suggest that the use of the Pentacam device could be an effective tool for monitoring the progression of AXL in myopic patients.

Author Contributions: Conceptualization, A.L.-M., B.G.-M., M.C.S.-G., R.C.-P., C.D.-H.-C., M.R.-L., J.-J.C.-D. and J.-M.S.-G.; methodology, A.L.-M., B.G.-M., M.C.S.-G., R.C.-P., C.D.-H.-C., M.R.-L., J.-J.C.-D. and J.-M.S.-G.; software, A.L.-M., B.G.-M., M.C.S.-G., R.C.-P., C.D.-H.-C., M.R.-L., J.-J.C.-D. and J.-M.S.-G.; validation, A.L.-M., B.G.-M., M.C.S.-G., R.C.-P., C.D.-H.-C., M.R.-L., J.-J.C.-D. and J.-M.S.-G.; formal analysis, A.L.-M., B.G.-M., M.C.S.-G., R.C.-P., C.D.-H.-C., M.R.-L., J.-J.C.-D. and J.-M.S.-G.; investigation, A.L.-M., B.G.-M., M.C.S.-G., R.C.-P., C.D.-H.-C., M.R.-L., J.-J.C.-D. and J.-M.S.-G.; resources, A.L.-M., B.G.-M., M.C.S.-G., R.C.-P., C.D.-H.-C., M.R.-L., J.-J.C.-D. and J.-M.S.-G.; data curation, A.L.-M., B.G.-M., M.C.S.-G., R.C.-P., C.D.-H.-C., M.R.-L., J.-J.C.-D. and J.-M.S.-G.; writing—original draft preparation, A.L.-M., B.G.-M., M.C.S.-G., R.C.-P., C.D.-H.-C., M.R.-L., J.-J.C.-D. and J.-M.S.-G.; writing—review and editing, A.L.-M., B.G.-M., M.C.S.-G., R.C.-P., C.D.-H.-C., M.R.-L., J.-J.C.-D. and J.-M.S.-G.; visualization, A.L.-M., B.G.-M., M.C.S.-G., R.C.-P., C.D.-H.-C., M.R.-L., J.-J.C.-D. and J.-M.S.-G.; supervision, A.L.-M., B.G.-M., M.C.S.-G., R.C.-P., C.D.-H.-C., M.R.-L., J.-J.C.-D. and J.-M.S.-G. All authors have read and agreed to the published version of the manuscript.

Funding: This research received no external funding.

Institutional Review Board Statement: The study was conducted in accordance with the Declaration of Helsinki and approved by the Ethics Committee of the University of Seville (0054-N-19 January 2019).

Informed Consent Statement: Informed consent was obtained from all subjects involved in the study.

Data Availability Statement: The data presented in this study are available on request from the corresponding author. The data are not publicly available due to their containing information that could compromise the privacy of research participants.

Acknowledgments: The authors appreciate the support offered by the members of the Department of Physics of Condensed Matter, Faculty of Physics, University of Seville, with special thanks to Javier Romero-Landa and Clara Conde-Amiano. In addition, the authors also appreciate the technical support offered by the members and facilities of the Faculty of Pharmacy, University of Seville, with special thanks to María Álvarez-de-Sotomayor.

Conflicts of Interest: The authors declare no conflict of interest. The funders had no role in the design of the study; in the collection, analyses, or interpretation of data; in the writing of the manuscript, or in the decision to publish the results.

References

1. Holden, B.A.; Fricke, T.R.; Wilson, D.A.; Jong, M.; Naidoo, K.S.; Sankaridurg, P.; Wong, T.Y.; Naduvilath, T.J.; Resnikoff, S. Global Prevalence of Myopia and High Myopia and Temporal Trends from 2000 through 2050. *Ophthalmology* **2016**, *123*, 1036–1042. [CrossRef] [PubMed]
2. Flitcroft, D.I.; He, M.; Jonas, J.B.; Jong, M.; Naidoo, K.; Ohno-Matsui, K.; Rahi, J.; Resnikoff, S.; Vitale, S.; Yannuzzi, L. IMI—Defining and classifying myopia: A proposed set of standards for clinical and epidemiologic studies. *Investig. Ophthalmol. Vis. Sci.* **2019**, *60*, M20–M30. [CrossRef]
3. Chua, S.Y.L.; Sabanayagam, C.; Cheung, Y.B.; Chia, A.; Valenzuela, R.K.; Tan, D.; Wong, T.Y.; Cheng, C.Y.; Saw, S.M. Age of onset of myopia predicts risk of high myopia in later childhood in myopic Singapore children. *Ophthalmic Physiol. Opt.* **2016**, *36*, 388–394. [CrossRef] [PubMed]
4. Young, T.L.; Metlapally, R.; Shay, A.E. Complex trait genetics of refractive error. *Arch. Ophthalmol.* **2007**, *125*, 38–48. [CrossRef]
5. Alvarez-Peregrina, C.C.; Sanchez-Tena, M.A.M.A.; Martinez-Perez, C.C.; Villa-Collar, C.C. Prevalence and Risk Factors of Myopia in Spain. *J. Ophthalmol.* **2019**, *2019*, 3419576. [CrossRef]
6. Charman, N. Myopia: Its prevalence, origins and control. *Ophthalmic Physiol. Opt.* **2011**, *31*, 3–6. [CrossRef]
7. Mutti, D.O.; Sinnott, L.T.; Mitchell, G.L.; Jones-Jordan, L.A.; Moeschberger, M.L.; Cotter, S.A.; Kleinstein, R.N.; Manny, R.E.; Twelker, J.D.; Zadnik, K. Relative peripheral refractive error and the risk of onset and progression of myopia in children. *Investig. Ophthalmol. Vis. Sci.* **2011**, *52*, 199–205. [CrossRef]
8. Mordechai, S.; Gradstein, L.; Pasanen, A.; Ofir, R.; El Amour, K.; Levy, J.; Belfair, N.; Lifshitz, T.; Joshua, S.; Narkis, G.; et al. High myopia caused by a mutation in LEPREL1, encoding prolyl 3-hydroxylase 2. *Am. J. Hum. Genet.* **2011**, *89*, 438–445. [CrossRef]
9. Martínez-Plaza, E.; Molina-Martín, A.; Arias-Puente, A.; Piñero, D.P. Clinical Validation of a New Optical Biometer for Myopia Control in a Healthy Pediatric Population. *Children* **2022**, *9*, 1713. [CrossRef]
10. Queirós, A.; Amorim-de-Sousa, A.; Fernandes, P.; Ribeiro-Queirós, M.S.; Villa-Collar, C.; González-Méijome, J.M. Mathematical Estimation of Axial Length Increment in the Control of Myopia Progression. *J. Clin. Med.* **2022**, *11*, 6200. [CrossRef]
11. Shammas, H.J.; Ortiz, S.; Shammas, M.C.; Kim, S.H.; Chong, C. Biometry measurements using a new large-coherence-length swept-source optical coherence tomographer. *J. Cataract Refract. Surg.* **2016**, *42*, 50–61. [CrossRef] [PubMed]
12. Monaghan, T.F.; Rahman, S.N.; Agudelo, C.W.; Wein, A.J.; Lazar, J.M.; Everaert, K.; Dmochowski, R.R. Foundational Statistical Principles in Medical Research: Sensitivity, Specificity, Positive Predictive Value, and Negative Predictive Value. *Medicina* **2021**, *57*, 503. [CrossRef] [PubMed]
13. Binney, N.; Hyde, C.; Bossuyt, P.M. On the Origin of Sensitivity and Specificity. *Ann. Intern. Med.* **2021**, *174*, 401–407. [CrossRef] [PubMed]
14. Sel, S.; Stange, J.; Kaiser, D.; Kiraly, L. Repeatability and agreement of Scheimpflug-based and swept-source optical biometry measurements. *Contact Lens Anterior Eye* **2017**, *40*, 318–322. [CrossRef]
15. Armstrong, R.A. Statistical guidelines for the analysis of data obtained from one or both eyes. *Ophthalmic Physiol. Opt.* **2013**, *33*, 7–14. [CrossRef]
16. Manna, P.; Karmakar, S.; Bhardwaj, G.K.; Mondal, A. Accommodative spasm and its different treatment approaches: A systematic review. *Eur. J. Ophthalmol.* **2022**. *Online ahead of print*. [CrossRef]
17. Jivrajka, R.; Shammas, M.C.; Boenzi, T.; Swearingen, M.; Shammas, H.J. Variability of axial length, anterior chamber depth, and lens thickness in the cataractous eye. *J. Cataract Refract. Surg.* **2008**, *34*, 289–294. [CrossRef]
18. Alfonso, J.F.; Ferrer-Blasco, T.; González-Méijome, J.M.; García-Manjarres, M.; Peixoto-de-Matos, S.C.; Montés-Micó, R. Pupil size, white-to-white corneal diameter, and anterior chamber depth in patients with myopia. *J. Refract. Surg.* **2010**, *26*, 891–898. [CrossRef]
19. Li, S.M.; Li, S.Y.; Kang, M.T.; Zhou, Y.H.; Li, H.; Liu, L.R.; Yang, X.Y.; Wang, Y.P.; Yang, Z.; Zhan, S.Y.; et al. Distribution of ocular biometry in 7- and 14-year-old Chinese children. *Optom. Vis. Sci.* **2015**, *92*, 566–572. [CrossRef]
20. Kato, K.; Kondo, M.; Takeuchi, M.; Hirano, K. Refractive error and biometrics of anterior segment of eyes of healthy young university students in Japan. *Sci. Rep.* **2019**, *9*, 15337. [CrossRef]
21. Zha, Y.; Feng, W.; Han, X.; Cai, J. Evaluation of myopic corneal diameter with the Orbscan II Topography System. *Graefe's Arch. Clin. Exp. Ophthalmol.* **2013**, *251*, 537–541. [CrossRef]
22. Li, S.M.; Iribarren, R.; Kang, M.T.; Li, H.; Li, S.Y.; Liu, L.R.; Sun, Y.Y.; Meng, B.; Zhan, S.Y.; Rozema, J.J.; et al. Corneal Power, Anterior Segment Length and Lens Power in 14-year-old Chinese Children: The Anyang Childhood Eye Study. *Sci. Rep.* **2016**, *6*, 20243. [CrossRef]
23. Bao, F.J.; Yu, A.Y.; Kassem, W.; Wang, Q.M.; Elsheikh, A. Biometry of the cornea in myopic Chinese patients. *J. Refract. Surg.* **2011**, *27*, 345–355. [CrossRef]
24. Logan, N.S.; Davies, L.N.; Mallen, E.A.H.; Gilmartin, B. Ametropia and ocular biometry in a U.K. university student population. *Optom. Vis. Sci.* **2005**, *82*, 261–266. [CrossRef]
25. Wei, S.; Sun, Y.; Li, S.; Hu, J.; Yang, X.; Lin, C.; Cao, K.; Du, J.; Guo, J.; Li, H.; et al. Refractive errors in university students in central China: The anyang university students eye study. *Investig. Ophthalmol. Vis. Sci.* **2018**, *59*, 4691–4700. [CrossRef]
26. Jorge, J.; Braga, A.; Queirós, A. Changes in myopia prevalence among first-year university students in 12 years. *Optom. Vis. Sci.* **2016**, *93*, 1262–1267. [CrossRef]

27. Shajari, M.; Cremonese, C.; Petermann, K.; Singh, P.; Müller, M.; Kohnen, T. Comparison of Axial Length, Corneal Curvature and Anterior Chamber Depth Measurements of 2 Recently Introduced Devices to a Known Biometer. *Am. J. Ophthalmol.* **2017**, *178*, 58–64. [CrossRef]
28. Trivedi, R.H.; Wilson, M.E. Globe axial length data in children using immersion A-scan ultrasound. *J. Cataract Refract. Surg.* **2021**, *47*, 1481–1482. [CrossRef]
29. Roessler, G.F.; Talab, Y.D.; Dietlein, T.S.; Dinslage, S.; Plange, N.; Walter, P.; Mazinani, B.A. Partial Coherence Laser Interferometry in Highly Myopic versus Emmetropic Eyes. *J. Ophthalmic Vis. Res.* **2014**, *9*, 169–173.
30. Chirapapaisan, C.; Srivannaboon, S.; Chonpimai, P. Efficacy of Swept-source Optical Coherence Tomography in Axial Length Measurement for Advanced Cataract Patients. *Optom. Vis. Sci. Off. Publ. Am. Acad. Optom.* **2020**, *97*, 186–191. [CrossRef]
31. Cooke, D.L.; Cooke, T.L. Approximating sum-of-segments axial length from a traditional optical low-coherence reflectometry measurement. *J. Cataract Refract. Surg.* **2019**, *45*, 351–354. [CrossRef]
32. Yu, J.; Wen, D.; Zhao, J.; Wang, Y.; Feng, K.; Wan, T.; Savini, G.; McAlinden, C.; Lin, X.; Niu, L.; et al. Comprehensive comparisons of ocular biometry: A network-based big data analysis. *Eye Vis.* **2022**, *10*, 1. [CrossRef] [PubMed]
33. Kanclerz, P.; Hoffer, K.J.; Bazylczyk, N.; Wang, X.; Savini, G. Optical Biometry and IOL Calculation in a Commercially Available Optical Coherence Tomography Device and Comparison with Pentacam AXL. *Am. J. Ophthalmol.* **2023**, *246*, 236–241. [CrossRef] [PubMed]
34. Moshirfar, M.; Tenney, S.; McCabe, S.; Schmid, G. Repeatability and reproducibility of the galilei G6 and its agreement with the pentacam®AXL in optical biometry and corneal tomography. *Expert Rev. Med. Devices* **2022**, *19*, 375–383. [CrossRef]
35. Chang, S.W.; Tsai, I.L.; Hu, F.R.; Lin, L.L.K.; Shih, Y.F. The cornea in young myopic adults. *Br. J. Ophthalmol.* **2001**, *85*, 916–920. [CrossRef] [PubMed]
36. Fang, Y.; Yokoi, T.; Nagaoka, N.; Shinohara, K.; Onishi, Y.; Ishida, T.; Yoshida, T.; Xu, X.; Jonas, J.B.; Ohno-Matsui, K. Progression of Myopic Maculopathy during 18-Year Follow-up. *Ophthalmology* **2018**, *125*, 863–877. [CrossRef] [PubMed]
37. Zadnik, K.; Mutti, D.O.; Friedman, N.E.; Qualley, P.A.; Jones, L.A.; Qiu, P.H.; Kim, H.S.; Hsu, J.C.; Moeschberger, M.L. Ocular predictors of the onset of juvenile myopia. *Investig. Ophthalmol. Vis. Sci.* **1999**, *40*, 1936–1943. [CrossRef]
38. Yebra-Pimentel, E.; González-Méijome, J.; Cerviño, A.; Giráldez, M.; González-Pérez, J.; Parafita, M. Asfericidad corneal en una población de adultos jóvenes: Implicaciones clínicas. *Arch. Soc. Esp. Oftalmol.* **2004**, *79*, 385–392. [CrossRef] [PubMed]
39. Su, D.H.W.; Wong, T.Y.; Foster, P.J.; Tay, W.T.; Saw, S.M.; Aung, T. Central Corneal Thickness and its Associations with Ocular and Systemic Factors: The Singapore Malay Eye Study. *Am. J. Ophthalmol.* **2009**, *147*, 709–716.e1. [CrossRef]
40. Zocher, M.T.; Rozema, J.J.; Oertel, N.; Dawczynski, J.; Wiedemann, P.; Rauscher, F.G. Biometry and visual function of a healthy cohort in Leipzig, Germany. *BMC Ophthalmol.* **2016**, *16*, 79. [CrossRef]
41. Hashemi, H.; Khabazkhoob, M.; Iribarren, R.; Emamian, M.H.; Fotouhi, A. Five-year change in refraction and its ocular components in the 40- to 64-year-old population of the Shahroud eye cohort study. *Clin. Exp. Ophthalmol.* **2016**, *44*, 669–677. [CrossRef] [PubMed]
42. Ampolu, N.; Yarravarapu, D.; Satgunam, P.; Varadharajan, L.S.; Bharadwaj, S.R. Impact of induced pseudomyopia and refractive fluctuations of accommodative spasm on visual acuity. *Clin. Exp. Optom.* **2022**. *Online ahead of print*. [CrossRef] [PubMed]

life

MDPI

Review

Complications of Small Aperture Intracorneal Inlays: A Literature Review

María Carmen Sánchez-González [1,*], Estanislao Gutiérrez-Sánchez [2], José-María Sánchez-González [1], Concepción De-Hita-Cantalejo [1], Ana-María Pinero-Rodríguez [2], Timoteo González-Cruces [3] and Raúl Capote-Puente [1]

1 Department of Physics of Condensed Matter, Optics Area, University of Seville, 41012 Seville, Spain
2 Department of Surgery, Ophthalmology Area, University of Seville, 41009 Seville, Spain
3 Department of Anterior Segment, Cornea and Refractive Surgery, Hospital La Arruzafa, 14012 Cordoba, Spain
* Correspondence: msanchez77@us.es

Abstract: Presbyopia can be defined as the refractive state of the eye in which, due to a physiological decrease in the ability to accommodate, it is not possible to sustain vision without fatigue in a prolonged manner, along with difficulty focusing near vision. It is estimated that its prevalence in 2030 will be approximately 2.1 billion people. Corneal inlays are an alternative in the correction of presbyopia. They are implanted beneath a laser-assisted in situ keratomileusis (LASIK) flap or in a pocket in the center of the cornea of the non-dominant eye. The purpose of this review is to provide information about intraoperative and postoperative KAMRA inlay complications in the available scientific literature. A search was conducted on PubMed, Web of Science, and Scopus with the following search strategy: ("KAMRA inlay" OR "KAMRA" OR "corneal inlay pinhole" OR "pinhole effect intracorneal" OR "SAICI" OR "small aperture intracorneal inlay") AND ("complication" OR "explantation" OR "explanted" OR "retired"). The bibliography consulted shows that the insertion of a KAMRA inlay is an effective procedure that improves near vision with a slight decrease in distance vision. However, postoperative complications such as corneal fibrosis, epithelial iron deposits, and stromal haze are described.

Keywords: small aperture intracorneal inlay; SAICI; KAMRA inlay; corneal inlay pinhole; intraoperative complications; postoperative complications; refractive surgery; presbyopia

Citation: Sánchez-González, M.C.; Gutiérrez-Sánchez, E.; Sánchez-González, J.-M.; De-Hita-Cantalejo, C.; Pinero-Rodríguez, A.-M.; González-Cruces, T.; Capote-Puente, R. Complications of Small Aperture Intracorneal Inlays: A Literature Review. *Life* **2023**, *13*, 312. https://doi.org/10.3390/life13020312

Academic Editor: Akio Oishi

Received: 11 January 2023
Revised: 20 January 2023
Accepted: 21 January 2023
Published: 22 January 2023

1. Introduction

Presbyopia can be defined as the refractive state of the eye in which, due to a physiological decrease in the ability to accommodate, it is not possible to sustain vision without fatigue in a prolonged manner, along with difficulty focusing on near vision [1–4]. Presbyopia depends not only on biological age but also on predominant factors such as visual defects and working distance and factors such as low light or fatigue disorders at the end of the day [5,6]. It is estimated that its prevalence in 2030 will be approximately 2.1 billion people [7].

Surgery for presbyopia includes a wide range of surgical approaches and procedures. Strategies include different corneal approaches, such as the application of excimer and femtosecond lasers in techniques such as monovision, sectoral modification in corneal multifocality (PresbyLASIK), and conductive keratoplasty, where electromagnetic radiation is used as the radiofrequency energy [8–10]. Similarly, corneal modifications assisted by femtosecond lasers such as IntraCOR and SUPRACOR are successfully used, where changes in the patterns of corneal aberrations are generated [11–13].

Currently, multifocal intraocular lenses (MIOLs) are also an effective alternative for the correction of pseudophakic presbyopia. MIOLs are grouped according to their optical characteristics: nonapodized (AcrySoft® IQ PanOptix®, Alcon Laboratories, Inc., Fort Worth, TX, USA) and apodized diffractive designs (ReSTOR®, Alcon Laboratories,

Inc., Fort Worth, TX, USA), and refractive (Rezoom^{TM}, AMO, Inc., Santa Ana, CA, USA), pseudo accommodative AcrySoft ReSTOR (Alcon Laboratories, Inc., Fort Worth, TX, USA), and extended depth of focus (EDOF) designs (Eden, SAV-IOLS.A., Neuchâtel, Switzerland) [14–16]. Additionally, correction of the multifocal design of the anterior chamber (Vivarte® Presbyopic, IOLtech, La Rochelle, France) with phakic intraocular lenses (pIOLs) is used for its predictability, fixation technique, and support for presbyopia correction [17,18].

KAMRA inlay surgery is a type of vision correction procedure that involves the implantation of a small, circular device called a KAMRA inlay into the cornea of the eye [19]. The KAMRA inlay is designed to help improve near vision in people with presbyopia, a condition that causes the loss of the eye's ability to focus on close objects as we age [20]. The KAMRA inlay is made of a thin circular disk of polymer material that is just a few millimeters in diameter. It is placed in a layer of the cornea called the stroma, which is located just beneath the outermost layer of the cornea, called the epithelium [21]. The inlay is placed in the non-dominant eye, which is typically the left eye for right-handed individuals. The KAMRA inlay works by creating a small, circular opening in the center of the cornea that allows light to pass through and focus on the retina at the back of the eye [22]. This opening, known as a pinhole, narrows the focus of light entering the eye and improves the eye's ability to see objects at close range. By using the "depth of focus" principle commonly used in photography, the KAMRA implant controls light transmission, allowing only focused light rays to reach the retina through a fixed 1.6 mm aperture. The inlay is designed to be small enough to be barely noticeable, but still large enough to allow a sufficient amount of light to pass through and improve near vision [23].

KAMRA inlay surgery is typically performed as an outpatient procedure and takes about 15–30 min to complete. The procedure is generally well tolerated and has a low risk of complications [24]. Before the procedure, the surgeon will numb the eye with anesthetic drops to reduce any discomfort during the procedure. The surgeon will then create a small flap in the cornea using a laser or a blade and lift it up to access the stroma [25]. Generally, KAMRA inlays are implanted mechanically or assisted by a femtosecond laser in people with presbyopia who do not have refractive errors. For people with refractive errors, a combined procedure, such as LASIK, can be performed to fix the refractive error at the same time as the inlay is implanted [26]. The KAMRA inlay is then carefully placed in the stroma and the corneal flap is replaced and sealed. After the procedure, the eye may be slightly red and swollen for a few days, and some people may experience mild discomfort or sensitivity to light [27]. These symptoms typically resolve on their own within a few days to a week. It is important for patients to follow the surgeon's instructions for postoperative care, which may include using eye drops or ointment to keep the eye moist and prevent infection [28]. The surgeon will also schedule follow-up appointments to monitor the healing process and ensure that the inlay is functioning properly. Most people who have KAMRA inlay surgery experience a significant improvement in their near vision within a few days to a week after the procedure [29]. The results of the procedure are generally long-lasting, although the inlay may need to be replaced after several years if the patient's vision begins to decline again [30]. KAMRA inlay surgery is an effective treatment option for people with presbyopia who want to improve their near vision without the need for glasses or contacts. It is generally a safe and well-tolerated procedure, with a low risk of complications. However, as with any surgical procedure, it is important for patients to carefully consider the potential risks and benefits and discuss them with their surgeon before deciding whether KAMRA inlay surgery is the right choice for them [27]. The approximate number of KAMRA inlays that have been implanted worldwide is 20.000 [31].

2. Corneal Inlay

Synthetic keratophakia was first described by Barraquer in 1949 [19]. However, the materials he used, i.e., flint glass and plexiglass, were found to be unsuitable due to biocompatibility issues. In 1960, other more transparent and permeable materials, hydrogel polymers, were tested to favor metabolic gradients through the stroma so that waste products pass into the aqueous humor and the flow of nutrients to the cornea is maintained [19].

There have been important advances in the design and material of corneal implants. Currently, the models used are thinner and permeable to oxygen, and the use of the femtosecond laser facilitates their intracorneal placement. Three types of implants are considered to have good results (Table 1): corneal reshaping inlays, Raindrop® Near Vision (ReVision Optics, Lake Forest, CA, USA) [32]; refractive inlays, Flexivue Microlens™ (Presbia Coöperatief U.A., Irvine, CA, USA) [27] and Icolens System™ (Neoptics AG, Hünenberg, Switzerland) [27]; and small aperture intracorneal inlays, KAMRA™ (AcuFocus Inc., Irvine, CA, USA) [20]. The Raindrop inlay was discontinued in 2018, just 2 years after its FDA approval in 2016, due to corneal haze, while Flexivue Microlens has been awaiting FDA approval in the United States since 2019 [33].

Table 1. Characteristics of corneal inlays.

Corneal Inlay	Name	Diameter (mm)	Thickness (μm)	Material	Placement (Flap/Pocket) μm
Reshaping	Raindrop®	1.5–2	Periphery (10) Center (30)	HAH	130–150
Refractive	Flexivue Microlens™	3,2	15–20	HAH with ultraviolet filter (Contaflex C126)	280–300
	Icolens™	3	15–20		
SAICI	KAMRA™	3,8 1.6 (Pinhole)	5	PFC	200–250

SAICI: small aperture intracorneal inlay; HAH: hydrophilic acrylic hydrogel; PFC: polyvinylidene fluoride and carbon.

The purpose of this review is to provide information about intraoperative and postoperative KAMRA inlay complications in the available scientific literature.

3. Literature Review

3.1. Search Strategy

A systematic literature search was performed using the PubMed/MEDLINE (77 articles), Web of Science (55 articles), and Scopus (54 articles) databases. The search strategy included the terms ("KAMRA inlay" OR "KAMRA" OR "corneal inlay pinhole" OR "pinhole effect intracorneal" OR "SAICI" OR "small aperture intracorneal inlay") AND ("complication" OR "explantation" OR "explanted" OR "retired"). Databases were searched for publications from January 2011 to January 2022.

The inclusion criteria were as follows: (1) studies with humans; (2) case reports; (3) case series; (4) cohort, cross-sectional, and case–control studies; and (5) randomized clinical trials. The exclusion criteria were as follows: (1) animal studies; (2) the article was a letter to the editor, conference abstract, study protocol, or literary review; (3) the article was not available in English; and (4) nonindexed publications.

A total of 186 articles were identified. After removing duplicates, article grading and data extraction were independently performed by two authors, MCSG and RCP, according to the inclusion and exclusion criteria. If there was a conflict with the selection of an article, the third author, JMSG, decided the outcome. A total of 20 articles were finally included in the review.

3.2. Results of Literature Search

The literature consulted shows that both the insertion of a KAMRA in presbyopic emmetropes [29,34–39] and insertion combined with LASIK [40–45] or PRK [46] in patients with ametropia, in addition to the insertion in presbyopic phakic patients [47], is an effective procedure that improves near vision with a slight decrease in distance vision in the eye with the implant or in both eyes.

Dexl et al. [37] followed up with 32 patients for five years. Their results showed a significant improvement in uncorrected near visual acuity (UNVA) after one year that remained stable up to month 36. At 60 months, UNVA decreased slightly. Similarly, this also occurred with uncorrected intermediate visual acuity (UIVA), which improved from 20/32 to 20/25 at 12 months, remained stable up to 36 months, and at 60 months decreased slightly to 20/32.

However, postoperative complications such as halos, dry eyes, and alterations in night vision have been described [42] and hyperopic regression has been described after a follow-up period [35,36,41,42,46]. There are authors who describe the appearance of corneal fibrosis [30,48], epithelial iron deposits [35], and keratocyte activation around the back surface of the inlay [49,50] as possible causes of haze that require explantation.

Although the KAMRA corneal inlay is a removable device, patients may experience residual corneal haze, hyperopic shift, and deficits in uncorrected distance visual acuity (UDVA) after explantation compared to pre-implantation UDVA [33].

Darian-Smith et al. retrospectively analyzed the visual outcomes of KAMRA inlay insertion in a cohort of patients reporting success of procedure, complications, patient satisfaction, and refractive outcomes at the TLC Laser Centre, Toronto. The explantation rate was 11.42%; 28.5% of patients required enhancements after inlay insertion [51]. Moshirfar et al. evaluated 10 years of KAMRA corneal inlay explantation and associated visual outcomes. KAMRA explantation rate was 8.2% across 10 years in Salt Lake City, Utah, USA [33].

3.3. *Complications*

3.3.1. Distance Visual Acuity

Although patients experience better near and intermediate vision after KAMRA surgery [51], follow-up with patients after implantation suggests decreased distance visual acuity in the implanted eye or in both eyes. Dexl et al. [35] reported a decrease in distance visual acuity with correction (CDVA) in emmetropic presbyopic patients. Additionally, Tomita et al. [44] and Vukich et al. [36] described how visual acuity decreased with distance in a group of post-LASIK and emmetropic patients after KAMRA implantation.

3.3.2. Refractive Changes

Refractive instability is also a predictable consequence after KAMRA surgery and has been reported by several authors [41,47,52]. Moshirfar et al. [53], in a case series of 50 patients, reported keratometric and topographic changes that led to a change in refraction. At 3 years, 54% of the eyes implanted with a KAMRA had a hyperopic manifest refractive spherical equivalent (MRSE), 40% were myopic with respect to the initial value, and the mean keratometry (K_m) was significantly increased at all postoperative measurements compared with baseline. Optimal near-vision results require slightly myopic MRSE, with −0.75 considered an ideal compromise between near and far vision.

3.3.3. Decentration

KAMRA inlay implantation has been described as successful in the majority of patients; in the actual scientific literature the decentration and repositioning rate ranges from 1.2% to 8.8% [33]. The small-aperture corneal inlay implantation technique creates an intrastromal pocket with a femtosecond laser and the pinhole device is later placed into this pocket via a small incision [53]. The criteria for KAMRA inlay placement are targeted in the center of the Purkinje reflex. In cases where the Purkinje reflex and the pupil are separated by a few microns, the KAMRA inlay needs to be placed between the Purkinje reflex and the pupil [48]. The exact placement of the pinhole inlay is necessary to achieve good visual and refractive results.

Decentration of the inlay implies damage to visual quality and poor refractive results [54]. In Figure 1, a nasal decentration of the KAMRA inlay observed using the section illumination of a biomicroscope is presented. Repositioning should be performed after implantation to achieve good refractive results and improve near, intermediate, and distance vision. It is important to note that the KAMRA inlay is placed while the patient is lying down and the effect of gravity pushes down on the KAMRA inlay, which could increase the decentration of the device in the intrastromal pocket [52].

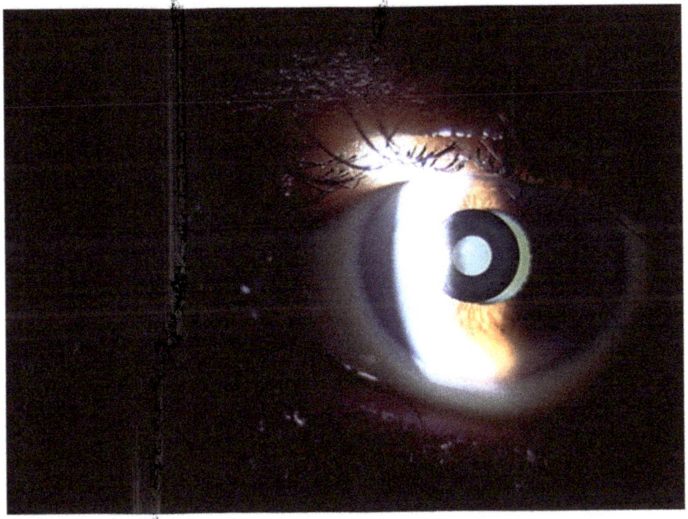

Figure 1. Slit-lamp examination. Decentration of a KAMRA inlay toward nasal area. A temporary area of the cornea free of the KAMRA is observed, which allows light to pass through.

New techniques and devices, such as ocular coherence tomography for anterior segment (AS-OCT) imaging, could help to locate and measure the decentration of a KAMRA inlay [50]. Figure 2 shows how AS-OCT technology reveals in a 3D-cube image how the inlay is partially displaced to the nasal zone. In another point of view, the AS-OCT could show us how deep the inlay is placed. Figure 3 shows the hyperreflective surface of the KAMRA inlay with a yellow and red image. Furthermore, the AS-OCT caliper could determine the depth of the inlay or the distance between the center of the pupil and the center of the inlay. This technology could be beneficial to perform after repositioning an inlay to move the pinhole device.

After the United States of America Food and Drug Administration approved KAMRA, the device gained in popularity [55]. The refractive results of this presbyopia treatment depend in large part on the centering of the pinhole device. Decentration of the KAMRA is principally due to the initial surgeon mispositioning it and inducing poor alignment [54]. This decentration is an intraoperative complication, while a displacement after the surgery indicates a migration in the inlay, therefore, migration is a postoperative complication. This issue has an explanation: the creation of a small pocket combined with the rough posterior face of the pinhole inlay implies that the device increased the adherence to the stromal pocket [41]. However, there are other intrastromal inlays that decentered after placement. The hydrogel corneal inlay Raindrop decentered after a corneal flap creation combined with a steroid that increased the intraocular pressure of the eye. In addition, the stromal keratopathy created by the increase in intraocular pressure possibly created an accumulation of fluid and a relocation of the hydrogel inlay [56].

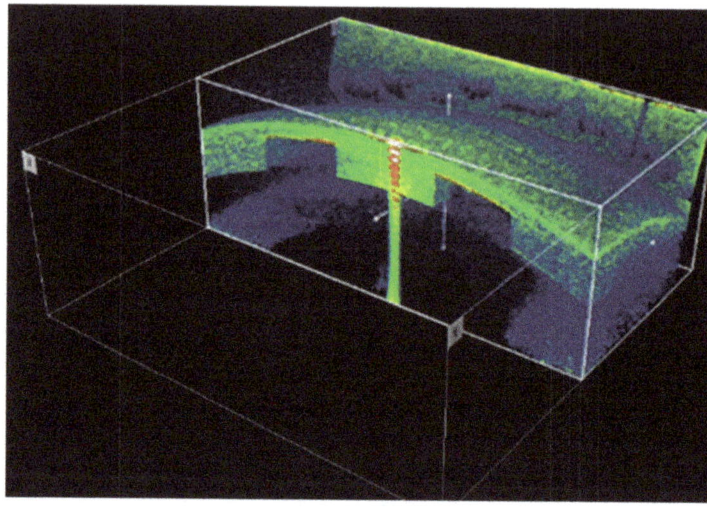

Figure 2. SD-AS-OCT of the anterior segment, revealing nasal decentration of the small-aperture corneal inlay. From left to right the nasal and temporal images are observed.

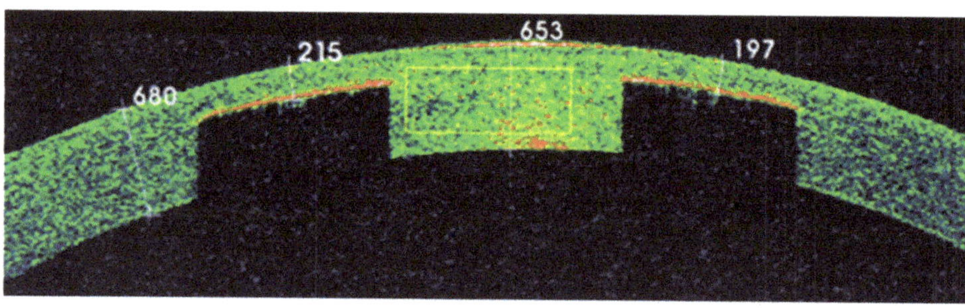

Figure 3. SD-AS-OCT of the anterior segment revealing a stromal hyperreflective signal. Caliper measurements from left to right (in μm): Total peripheral pachymetry (680), nasal KAMRA pocket depth (215), total central pachymetry (653), temporal KAMRA pocket depth (197).

3.3.4. Migration and Extrusion

Migration and extrusion are two potential complications that can occur with the KAMRA inlay [24,28,40,49,52]. Migration refers to the movement of the inlay within the cornea after it has been implanted. This can occur if the inlay becomes dislodged from its intended position within the cornea. Extrusion refers to the complete removal of the inlay from the cornea [30,41,57]. Both migration and extrusion can occur as a result of improper implantation, infection, inflammation, or other factors. Symptoms of migration or extrusion may include blurred vision, eye irritation, and discomfort. In severe cases, migration or extrusion of the KAMRA inlay may require surgical intervention to remove the device and restore vision [24,28,40,49,52].

The permeability of implants to water and other molecules (especially glucose) determines the biocompatibility of the implant and its survival in the corneal stroma [56,58]. Extrusion is usually preceded by stromal necrosis. Initially, inlays were designed of poly (methyl methacrylate) and polysulfone. The low permeability of these materials caused inadequate corneal nutrition, inducing thinning of the anterior stroma and keratolysis [19]. Hydrogel implants with variable water content have shown good results and absence of long-term extrusion.

3.3.5. Stromal Haze

Corneal haze (Figure 4) is a common complication of KAMRA inlay surgery, although it is typically mild and does not cause significant vision loss. Corneal haze is a condition in which the cornea becomes cloudy or hazy, which can interfere with vision [27]. It is caused by the formation of scar tissue in the cornea as a result of the surgical procedure [25]. Symptoms of corneal haze may include blurred or hazy vision, glare or halos around lights, and sensitivity to light. These symptoms may be more noticeable at night or in low light conditions. In most cases, corneal haze is mild and does not cause significant vision loss, but it can sometimes interfere with activities such as driving or reading [59,60]. Treatment for corneal haze typically involves the use of eye drops or ointments to reduce inflammation and promote healing. In some cases, laser treatment may be necessary to remove the scar tissue and improve vision [30,48]. Most people with corneal haze experience improvement in their vision within a few weeks to a few months after treatment, although it may take longer for some people. It is important for people who have had KAMRA inlay surgery to be aware of the potential for corneal haze and to monitor their vision closely after the procedure. Overall, while corneal haze is a potential complication of KAMRA inlay surgery, it is generally mild and can be effectively treated with the use of eye drops or ointments and, in some cases, laser treatment [27,40,48].

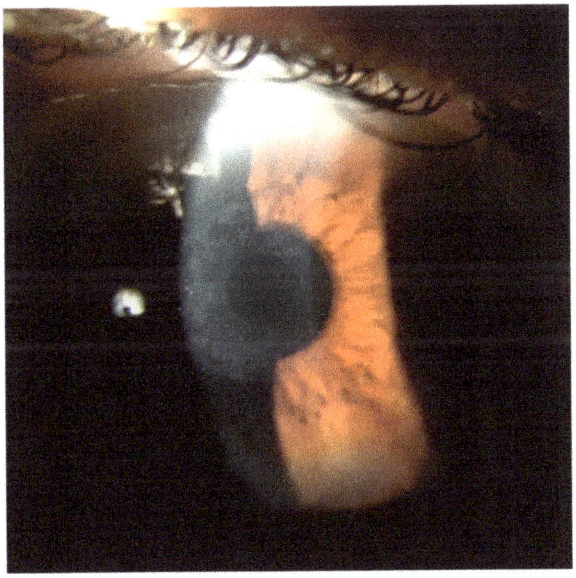

Figure 4. Slit-lamp examination at 2 months after removal. Stromal leukoma-shaped 360 degree ring (stromal footprint) associated with the KAMRA and corneal epithelial iron deposits in a half-moon shape (similar to a Fleischer ring).

The migration of surface proinflammatory cytokines is a local source of proteolytic enzymes and cytokines responsible for interface haze formation [57,61]. It is a predictable complication in response to the implantation of synthetic materials [61]. The greater the depth of the inlay implantation, the lower the keratocytic activation and development of haze [41]. Optical coherence tomography (OCT) is very effective in guiding the depth of femtosecond laser pocket creation for inlay implantation.

The cornea is an avascular tissue: on its anterior face, in intimate contact with the precorneal tear film, it absorbs oxygen, and on its posterior face, bathed by aqueous humor by diffusion, it receives glucose.

Keratocytes occupy 3–5% of the stromal volume. Its function consists of maintaining the collagen fibers and the extracellular matrix through a constant synthesis activity favoring metabolic gradients through the stroma so that waste products pass into the aqueous humor and nutrients flow to the cornea [62].

The microperforations of the KAMRA inlay optimize the flow of nutrients from the cornea [63]. However, the small-aperture corneal inlay can disrupt the exchange of oxygen and glucose. The result is the release of growth factors and proinflammatory cytokines that can generate myofibroblasts and be the cause of persistent fibrosis [30,48].

Numerous authors describe a series of cases of stromal opacity after implantation of a KAMRA [38,40,45,60]. The treatment of severe opacity includes the use of topically applied drugs. Corticosteroids and mitomycin C are usually effective. If the opacity persists over time, implant explantation is recommended [50].

3.3.6. Infectious Keratitis

Infectious keratitis is an infection of the cornea; it is a serious and potentially vision-threatening condition that can occur after KAMRA inlay surgery, although it is rare [48]. Infectious keratitis is caused by bacteria, fungi, or viruses that enter the eye and infect the cornea. It can occur as a result of a variety of factors, including trauma to the eye, contact lens wear, and surgery [26,64]. KAMRA inlay surgery involves the creation of a small flap in the cornea, which can increase the risk of infection if the eye is not properly cared for after the procedure. Symptoms of infectious keratitis may include redness, pain, and sensitivity to light in the affected eye, as well as discharge around the eye [26,64]. In some cases, the infection may cause the cornea to become cloudy or hazy, which can interfere with vision. Treatment for infectious keratitis typically involves the use of antibiotics or antifungal medications to clear the infection and prevent further damage to the eye. In severe cases, surgery may be necessary to remove the infected tissue and prevent the spread of the infection [28,48].

Duignan et al. [28] describe a series of cases that, after implantation, presented anterior-chamber cellular reactions, epithelial defects, conjunctival hyperemia, and corneal infiltrates characteristic of infectious keratitis. The patients were treated with antibiotics. Bouheraoua et al. [65] described the case of a woman with epithelial growth around an implant to correct hyperopia (Permavision, Anamed Inc., Lake Forest, CA, USA). It was probably the cause of the infectious keratitis that appeared seven years after the surgery.

3.3.7. Epithelial Ingrowth

Epithelial ingrowth is a potential complication that can occur after the implantation of a KAMRA inlay [61]. Epithelial ingrowth occurs when cells from the epithelium, the outer layer of the cornea, grow into the area surrounding the inlay [26]. Due to poor pocket adhesion or the presence of epithelial foreign bodies, cells migrate from the stromal pocket edge toward the pocket–stromal interface [26]. This can cause the inlay to become dislodged or distorted, leading to reduced vision and other symptoms [59]. Epithelial ingrowth may also cause inflammation and scarring, which can further compromise vision [66]. Epithelial ingrowth is more likely to occur if the inlay is not properly implanted or if there is damage to the cornea during the implantation procedure. It can also occur as a result of infection or inflammation. Symptoms of epithelial ingrowth may include blurred vision, eye irritation, and discomfort. In severe cases, surgical intervention may be necessary to remove the inlay and restore vision [61].

Dexl et al. [35] described a case of epithelial growth in a series of 32 cases with a KAMRA inlay. The patient presented epithelial growth after lifting the flap for the treatment of postoperative striae. Rafic et al. [59] describe epithelial growth in a 52-year-old man with basement membrane dystrophy after undergoing a combination of hyperopic laser keratomileusis and KAMRA implantation. Most likely, epithelial growth could be the cause of the basement membrane dystrophy that could have leaked proinflammatory cytokines and chemokines to the stromal interface and amplified keratocyte activation.

3.3.8. Binocular Vision

Castro et al. [67] showed a deterioration of binocular vision in a group of patients who underwent a simulation of KAMRA inlay. Anisocoria was induced with a contact lens that has a partially opaque peripheral area. The lens was placed on the nondominant eye. Measurements were made under two conditions: induced anisocoria and induced anisocoria combined with monovision using two additional powers inserted in a trial frame: +0.75 and +1.25. Stereoacuity was performed at three distances: near, intermediate, and far. The study showed a deterioration in stereoacuity in all induced anisocoria conditions. This deterioration was significant at intermediate and close distances. Additionally, Lin et al. [68] reported a deterioration in stereo acuity after KAMRA surgery.

Binocularity is the result of the processing of motor and sensory skills that allows obtaining a spatial reference from using both eyes simultaneously, merging a single image at the cortical level; when this happens, a stereoscopic vision or three-dimensional image is achieved [69–72]. Any type of alteration in binocular vision will establish a lack of brain integration and interpretation.

4. Conclusions

KAMRA inlay surgeries in presbyopic emmetropes and KAMRA inlay surgeries combined with LASIK in patients with ametropia is an effective procedure that improves uncorrected near visual acuity with a slight decrease in distance vision in the eye with the implant or in both eyes.

This pinhole implant surgery can have moderate to severe complications, such as refractive instability, epithelial ingrowth, or infectious keratitis. Refractive outcomes should be considered with caution due to possible conflicts of interest between the authors of the publications and manufacturers.

Author Contributions: Conceptualization, M.C.S.-G. and R.C.-P.; methodology, A.-M.P.-R. and E.G.-S.; software, C.D.-H.-C.; validation, J.-M.S.-G. and T.G.-C.; formal analysis, J.-M.S.-G.; investigation, A.-M.P.-R. and E.G.-S.; resources, C.D.-H.-C.; data curation, T.G.-C.; writing—original draft preparation, M.C.S.-G. and R.C.-P.; writing—review and editing, J.-M.S.-G. and T.G.-C.; visualization, C.D.-H.-C.; supervision, J.-M.S.-G.; project administration, N/A (Not Applicable); funding acquisition, N/A. All authors have read and agreed to the published version of the manuscript.

Funding: The English edition has been subsidized with the help of emerging groups from the 7th Own Research Plan of the University of Seville.

Acknowledgments: The authors acknowledge the support offered by the members of the Pinero Ophthalmology Clinic and Capote Vision Center.

Conflicts of Interest: The authors declare no conflict of interest.

References

1. Kollbaum, P.S.; Bradley, A. Correction of presbyopia: Old problems with old (and new) solutions. *Clin. Exp. Optom.* **2020**, *103*, 21–30. [CrossRef]
2. Wolffsohn, J.S.; Davies, L.N. Presbyopia: Effectiveness of correction strategies. *Prog. Retin. Eye Res.* **2019**, *68*, 124–143. [CrossRef]
3. Ayoub, S.C.; Ahmad, M. Presbyopia: Clinical Update. *Insight* **2017**, *42*, 29–36.
4. Charman, W.N. Virtual Issue Editorial: Presbyopia—Grappling with an age-old problem. *Ophthalmic Physiol. Opt.* **2017**, *37*, 655–660. [CrossRef]
5. Xu, R.; Gil, D.; Dibas, M.; Hare, W.; Bradley, A. The Effect of Light Level and Small Pupils on Presbyopic Reading Performance. *Investig. Ophthalmol. Vis. Sci.* **2016**, *57*, 5656–5664. [CrossRef]
6. Guillon, M.; Dumbleton, K.; Theodoratos, P.; Gobbe, M.; Wooley, C.B.; Moody, K. The Effects of Age, Refractive Status, and Luminance on Pupil Size. *Optom. Vis. Sci. Off. Publ. Am. Acad. Optom.* **2016**, *93*, 1093–1100. [CrossRef]
7. Fricke, T.R.; Tahhan, N.; Resnikoff, S.; Papas, E.; Burnett, A.; Ho, S.M.; Naduvilath, T.; Naidoo, K.S. Global Prevalence of Presbyopia and Vision Impairment from Uncorrected Presbyopia: Systematic Review, Meta-analysis, and Modelling. *Ophthalmology* **2018**, *125*, 1492–1499. [CrossRef]
8. Alió, J.L.; Chaubard, J.J.; Caliz, A.; Sala, E.; Patel, S. Correction of presbyopia by technovision central multifocal LASIK (presbyLASIK). *J. Refract. Surg.* **2006**, *22*, 453–460. [CrossRef]

9. Pallikaris, I.G.; Panagopoulou, S.I. PresbyLASIK approach for the correction of presbyopia. *Curr. Opin. Ophthalmol.* **2015**, *26*, 265–272. [CrossRef]
10. Gil-Cazorla, R.; Shah, S.; Naroo, S.A. A review of the surgical options for the correction of presbyopia. *Br. J. Ophthalmol.* **2016**, *100*, 62–70. [CrossRef]
11. Holzer, M.P.; Knorz, M.C.; Tomalla, M.; Neuhann, T.M.; Auffarth, G.U. Intrastromal femtosecond laser presbyopia correction: 1-year results of a multicenter study. *J. Refract. Surg.* **2012**, *28*, 182–188. [CrossRef] [PubMed]
12. Khoramnia, R.; Fitting, A.; Rabsilber, T.M.; Thomas, B.C.; Auffarth, G.U.; Holzer, M.P. Intrastromal femtosecond laser surgical compensation of presbyopia with six intrastromal ring cuts: 3-year results. *Br. J. Ophthalmol.* **2015**, *99*, 170–176. [CrossRef] [PubMed]
13. Sánchez-González, J.-M.; Alonso-Aliste, F.; Amián-Cordero, J.; Sánchez-González, M.C.; De-Hita-Cantalejo, C. Refractive and Visual Outcomes of SUPRACOR TENEO 317 LASIK for Presbyopia in Hyperopic Eyes: 24-Month Follow-up. *J. Refract. Surg.* **2019**, *35*, 591–598. [CrossRef] [PubMed]
14. Martinez-Perez, C.; Alvarez-Peregrina, C.; Villa-Collar, C.; Arance-Gil, A.; Sánchez-Tena, M.A. Citation network analysis of the various types of multifocal intraocular lenses. *Arch. Soc. Esp. Oftalmol.* **2021**, *96*, 527–544. [CrossRef]
15. Rampat, R.; Gatinel, D. Multifocal and Extended Depth-of-Focus Intraocular Lenses in 2020. *Ophthalmology* **2021**, *128*, e164–e185. [CrossRef]
16. Barnett, B.P. FOCUSED (Femtosecond Optimized Continuous Uncorrected Sight with EDOF and Diffractive Multifocal IOLs)—A Review. *Curr. Opin. Ophthalmol.* **2021**, *32*, 3–12. [CrossRef]
17. Alió, J.L.; Mulet, M.E. Presbyopia correction with an anterior chamber phakic multifocal intraocular lens. *Ophthalmology* **2005**, *112*, 1368–1374. [CrossRef]
18. Martínez-Plaza, E.; López-Miguel, A.; Holgueras, A.; Barraquer, R.I.; Alió, J.L.; Maldonado, M.J. Phakic intraocular lenses: Recent advances and innovations. *Arch. Soc. Esp. Oftalmol.* **2020**, *95*, 178–187. [CrossRef]
19. Waring, G.O.; Klyce, S.D. Corneal inlays for the treatment of presbyopia. *Int. Ophthalmol. Clin.* **2011**, *51*, 51–62. [CrossRef]
20. Lindstrom, R.L.; MacRae, S.M.; Pepose, J.S.; Hoopes, P.C. Corneal inlays for presbyopia correction. *Curr. Opin. Ophthalmol.* **2013**, *24*, 281–287. [CrossRef]
21. Seyeddain, O.; Riha, W.; Hohensinn, M.; Nix, G.; Dexl, A.K.; Grabner, G. Refractive surgical correction of presbyopia with the acufocus small aperture corneal inlay: Two-year follow-up. *J. Refract. Surg.* **2010**, *26*, 707–715. [CrossRef]
22. Dexl, A.K.; Seyeddain, O.; Riha, W.; Hohensinn, M.; Hitzl, W.; Grabner, G. Reading performance after implantation of a small-aperture corneal inlay for the surgical correction of presbyopia: Two-year follow-up. *J. Cataract Refract. Surg.* **2011**, *37*, 525–531. [CrossRef] [PubMed]
23. Langenbucher, A.; Goebels, S.; Szentmáry, N.; Seitz, B.; Eppig, T. Vignetting and field of view with the KAMRA corneal inlay. *Biomed Res. Int.* **2013**, *2013*, 154593. [CrossRef] [PubMed]
24. Sánchez-González, J.M.; Borroni, D.; Rachwani-Anil, R.; Rocha-de-Lossada, C. Refractive corneal inlay implantation outcomes: A preliminary systematic review. *Int. Ophthalmol.* **2022**, *42*, 713–722. [CrossRef]
25. Pluma-Jaramago, I.; Rocha-de-Lossada, C.; Rachwani-Anil, R.; Sánchez-González, J.M. Small-aperture intracorneal inlay implantation in emmetropic presbyopic patients: A systematic review. *Eye* **2022**, *36*, 1747–1753. [CrossRef] [PubMed]
26. Ting, D.S.J.; Srinivasan, S.; Danjoux, J.P. Epithelial ingrowth following laser in situ keratomileusis (LASIK): Prevalence, risk factors, management and visual outcomes. *BMJ Open Ophthalmol.* **2018**, *3*, e000133. [CrossRef]
27. Fenner, B.J.; Moriyama, A.S.; Mehta, J.S. Inlays and the cornea. *Exp. Eye Res.* **2021**, *205*, 108474. [CrossRef]
28. Duignan, E.S.; Farrell, S.; Treacy, M.P.; Fulcher, T.; O'Brien, P.; Power, W.; Murphy, C.C. Corneal inlay implantation complicated by infectious keratitis. *Br. J. Ophthalmol.* **2016**, *100*, 269–273. [CrossRef]
29. Seyeddain, O.; Hohensinn, M.; Riha, W.; Nix, G.; Rückl, T.; Grabner, G.; Dexl, A.K. Small-aperture corneal inlay for the correction of presbyopia: 3-year follow-up. *J. Cataract Refract. Surg.* **2012**, *38*, 35–45. [CrossRef]
30. Ong, H.S.; Chan, A.S.; Yau, C.W.; Mehta, J.S. Corneal inlays for presbyopia explanted due to corneal haze. *J. Refract. Surg.* **2018**, *34*, 357–360. [CrossRef]
31. Moshirfar, M.; Henrie, M.K.; Payne, C.J.; Ply, B.K.; Ronquillo, Y.C.; Linn, S.H.; Hoopes, P.C. Review of Presbyopia Treatment with Corneal Inlays and New Developments. *Clin. Ophthalmol.* **2022**, *16*, 2781–2795. [CrossRef] [PubMed]
32. Garza, E.B.; Gomez, S.; Chayet, A.; Dishler, J. One-year safety and efficacy results of a hydrogel inlay to improve near vision in patients with emmetropic presbyopia. *J. Refract. Surg.* **2013**, *29*, 166–172. [CrossRef]
33. Moshirfar, M.; Lau, C.K.; Chartrand, N.A.; Parsons, M.T.; Stapley, S.; Bundogji, N.; Ronquillo, Y.C.; Linn, S.H.; Hoopes, P.C. Explantation of KAMRA Corneal Inlay: 10-Year Occurrence and Visual Outcome Analysis. *Clin. Ophthalmol.* **2022**, *16*, 3327–3337. [CrossRef]
34. Waring IV, G.O. Correction of presbyopia with a small aperture corneal inlay. *J. Refract. Surg.* **2011**, *27*, 842–845. [CrossRef] [PubMed]
35. Dexl, A.K.; Jell, G.; Strohmaier, C.; Seyeddain, O.; Riha, W.; Rückl, T.; Bachernegg, A.; Grabner, G. Long-term outcomes after monocular corneal inlay implantation for the surgical compensation of presbyopia. *J. Cataract Refract. Surg.* **2015**, *41*, 566–575. [CrossRef] [PubMed]

36. Vukich, J.A.; Durrie, D.S.; Pepose, J.S.; Thompson, V.; van de Pol, C.; Lin, L. Evaluation of the small-aperture intracorneal inlay: Three-year results from the cohort of the U.S. Food and Drug Administration clinical trial. *J. Cataract Refract. Surg.* **2018**, *44*, 541–556. [CrossRef] [PubMed]
37. Dexl, A.K.; Seyeddain, O.; Riha, W.; Hohensinn, M.; Rückl, T.; Reischl, V.; Grabner, G. One-year visual outcomes and patient satisfaction after surgical correction of presbyopia with an intracorneal inlay of a new design. *J. Cataract Refract. Surg.* **2012**, *38*, 262–269. [CrossRef] [PubMed]
38. Seyeddain, O.; Bachernegg, A.; Riha, W.; Rückl, T.; Reitsamer, H.; Grabner, G.; Dexl, A.K. Femtosecond laser-assisted small-aperture corneal inlay implantation for corneal compensation of presbyopia: Two-year follow-up. *J. Cataract Refract. Surg.* **2013**, *39*, 234–241. [CrossRef]
39. Moshirfar, M.; Quist, T.S.; Skanchy, D.F.; Wallace, R.T.; Linn, S.H.; Hoopes, P.C. Six-month visual outcomes for the correction of presbyopia using a small-aperture corneal inlay: Single-site experience. *Clin. Ophthalmol.* **2016**, *10*, 2191–2198. [CrossRef]
40. Igras, E.; O'Caoimh, R.; O'Brien, P.; Power, W. Long-term results of combined LASIK and monocular small-aperture corneal inlay implantation. *J. Refract. Surg.* **2016**, *32*, 379–384. [CrossRef]
41. Fattah, M.A.; Mehanna, C.J.; Antonios, R.; Abiad, B.; Jabbur, N.S.; Awwad, S.T. Five-year results of combined small- A perture corneal inlay implantation and lasik for the treatment of hyperopic presbyopic eyes. *J. Refract. Surg.* **2020**, *36*, 498–505. [CrossRef] [PubMed]
42. Tomita, M.; Kanamori, T.; Waring IV, G.O.; Yukawa, S.; Yamamoto, T.; Sekiya, K.; Tsuru, T. Simultaneous corneal inlay implantation and laser in situ keratomileusis for presbyopia in patients with hyperopia, myopia, or emmetropia: Six-month results. *J. Cataract Refract. Surg.* **2012**, *38*, 495–506. [CrossRef] [PubMed]
43. Tomita, M.; Waring, G.O. One-year results of simultaneous laser in situ keratomileusis and small-aperture corneal inlay implantation for hyperopic presbyopia: Comparison by age. *J. Cataract Refract. Surg.* **2015**, *41*, 152–161. [CrossRef] [PubMed]
44. Tomita, M.; Kanamori, T.; Waring IV, G.O.; Nakamura, T.; Yukawa, S. Small-aperture corneal inlay implantation to treat presbyopia after laser in situ keratomileusis. *J. Cataract Refract. Surg.* **2013**, *39*, 898–905. [CrossRef]
45. Jalali, S.; Aus Der Au, W.; Shaarawy, T. AcuFocus Corneal Inlay to Correct Presbyopia Using Femto-LASIK. One Year Results of a Prospective Cohort Study. *Klin. Monbl. Augenheilkd.* **2016**, *233*, 360–364. [CrossRef] [PubMed]
46. Moshirfar, M.; Bean, A.E.; Albarracin, J.C.; Rebenitsch, R.L.; Wallace, R.T.; Birdsong, O.C. Retrospective comparison of visual outcomes after KAMRA corneal inlay implantation with simultaneous PRK or LASIK. *J. Refract. Surg.* **2018**, *34*, 310–315. [CrossRef]
47. Ylmaz, Ö.F.; Alagöz, N.; Pekel, G.; Azman, E.; Aksoy, E.F.; Çakr, H.; Bozkurt, E.; Demirok, A. Intracorneal inlay to correct presbyopia: Long-term results. *J. Cataract Refract. Surg.* **2011**, *37*, 1275–1281. [CrossRef]
48. Romito, N.; Basli, E.; Goemaere, I.; Borderie, V.; Laroche, L.; Bouheraoua, N. Persistent corneal fibrosis after explantation of a small-aperture corneal inlay. *J. Cataract Refract. Surg.* **2019**, *45*, 367–371. [CrossRef]
49. Abbouda, A.; Javaloy, J.; Alió, J.L. Confocal microscopy evaluation of the corneal response following AcuFocus KAMRA inlay implantation. *J. Refract. Surg.* **2014**, *30*, 172–178. [CrossRef]
50. Tomita, M.; Huseynova, T. Evaluating the short-term results of KAMRA inlay implantation using real-time optical coherence tomography-guided femtosecond laser technology. *J. Refract. Surg.* **2014**, *30*, 326–329. [CrossRef]
51. Darian-Smith, E.; Gouvea, L.; Gendler, S.; Alshaker, S.; Din, N.; Weill, Y.; Skouras, N.; Rabinovitch, T.; Singal, N.; Chan, C.C.; et al. KAMRA presbyopic inlay refractive outcomes: A Canadian perspective. *Can. J. Ophthalmol.* **2022**. [CrossRef] [PubMed]
52. Amigó, A.; Martinez-Sorribes, P.; Recuerda, M. Late-onset refractive shift after small-aperture corneal inlay implantation. *J. Cataract Refract. Surg.* **2018**, *44*, 658–664. [CrossRef] [PubMed]
53. Moshirfar, M.; Desautels, J.D.; Walker, B.D.; Birdsong, O.C.; Skanchy, D.F.; Quist, T.S.; Murri, M.S.; Linn, S.H.; Hoopes, P.C.; Hoopes, P.C. Long-term changes in keratometry and refraction after small aperture corneal inlay implantation. *Clin. Ophthalmol.* **2018**, *12*, 1931–1938. [CrossRef] [PubMed]
54. Hoopes, P.C.; Walker, B.D.; Birdsong, O.C.; Moshirfar, M. Small-aperture corneal inlay repositioning. *J. Cataract Refract. Surg.* **2018**, *44*, 3–5. [CrossRef]
55. Hoopes, P.C.; Desautels, J.D.; Moshirfar, M.; Linn, S.H.; Mamalis, N. Neodymium:YAG laser posterior capsulotomy in eye with an intrastromal inlay. *J. Cataract Refract. Surg.* **2017**, *43*, 699–702. [CrossRef]
56. Pinsky, P.M. Three-dimensional modeling of metabolic species transport in the cornea with a hydrogel intrastromal inlay. *Investig. Ophthalmol. Vis. Sci.* **2014**, *55*, 3093–3106. [CrossRef]
57. Sánchez-González, J.M.; Amián-Cordero, J.; Alonso-Aliste, F. Permavision intracorneal inlay after sixteen years. Regression of initial refractive hyperopia. *Contact Lens Anterior Eye* **2020**, *43*, 512–514. [CrossRef]
58. Larrea, X.; De Courten, C.; Feingold, V.; Burger, J.; Büchler, P. Oxygen and glucose distribution after intracorneal lens implantation. *Optom. Vis. Sci.* **2007**, *84*, 1074–1081. [CrossRef]
59. Antonios, R.; Jabbur, N.S.; Ahmed, M.A.; Awwad, S.T. Refractory interface haze developing after epithelial ingrowth following laser in situ keratomileusis and small aperture corneal inlay implantation. *Am. J. Ophthalmol. Case Reports* **2018**, *10*, 10–12. [CrossRef]
60. Fenner, B.J.; Liu, Y.C.; Koh, S.K.; Gao, Y.; Deng, L.; Beuerman, R.W.; Zhou, L.; Theng, J.T.S.; Mehta, J.S. Mediators of corneal haze following implantation of presbyopic corneal inlays. *Investig. Ophthalmol. Vis. Sci.* **2019**, *60*, 868–876. [CrossRef]

61. Riau, A.K.; Liu, Y.C.; Yam, G.H.F.; Mehta, J.S. Stromal keratophakia: Corneal inlay implantation. *Prog. Retin. Eye Res.* **2020**, *75* 100780. [CrossRef]
62. Petroll, W.M.; Miron-Mendoza, M. Mechanical interactions and crosstalk between corneal keratocytes and the extracellular matrix *Exp. Eye Res.* **2015**, *133*, 49–57. [CrossRef] [PubMed]
63. Moarefi, M.A.; Bafna, S.; Wiley, W. A Review of Presbyopia Treatment with Corneal Inlays. *Ophthalmol. Ther.* **2017**, *6*, 55–65 [CrossRef] [PubMed]
64. Konstantopoulos, A.; Kuo, J.; Anderson, D.; Hossain, P. Assessment of the Use of Anterior Segment Optical Coherence Tomography in Microbial Keratitis. *Am. J. Ophthalmol.* **2008**, *146*, 534–542. [CrossRef]
65. Bouheraoua, N.; Caillaux, V.; Sandali, O.; Laroche, L.; Borderie, V.M. Acanthamoeba keratitis associated with intracorneal hydrogel inlay. *J. Fr. Ophtalmol.* **2016**, *39*, e37–e41. [CrossRef] [PubMed]
66. Yesilirmak, N.; Chhadva, P.; Cabot, F.; Galor, A.; Yoo, S.H. Post-Laser in Situ Keratomileusis Epithelial Ingrowth: Treatment, Recurrence, and Long-Term Results. *Cornea* **2018**, *37*, 1517–1521. [CrossRef]
67. Castro, J.J.; Ortiz, C.; Jiménez, J.R.; Ortiz-Peregrina, S.; Casares-López, M. Stereopsis simulating small-aperture corneal inlay and monovision conditions. *J. Refract. Surg.* **2018**, *34*, 482–488. [CrossRef] [PubMed]
68. Linn, S.H.; Skanchy, D.F.; Quist, T.S.; Desautels, J.D.; Moshirfar, M. Stereoacuity after small aperture corneal inlay implantation. *Clin. Ophthalmol.* **2017**, *11*, 233–235. [CrossRef]
69. Han, S.B.; Yang, H.K.; Kim, J.; Hong, K.; Lee, B.; Hwang, J.-M. Quantification of Stereopsis in Patients with Impaired Binocularity. *Optom. Vis. Sci. Off. Publ. Am. Acad. Optom.* **2016**, *93*, 588–593. [CrossRef]
70. Rose, T.; Bonhoeffer, T. Experience-dependent plasticity in the lateral geniculate nucleus. *Curr. Opin. Neurobiol.* **2018**, *53*, 22–28. [CrossRef]
71. Levi, D.M. Rethinking amblyopia 2020. *Vision Res.* **2020**, *176*, 118–129. [CrossRef] [PubMed]
72. Ooi, T.L.; He, Z.J. Sensory Eye Dominance: Relationship Between Eye and Brain. *Eye Brain* **2020**, *12*, 25–31. [CrossRef] [PubMed]

life

Review

Ophthalmological Approach for the Diagnosis of Dry Eye Disease in Patients with Sjögren's Syndrome

Robinson T. Barrientos [1], Fernando Godín [2], Carlos Rocha-De-Lossada [3,4,5,6], Matias Soifer [7], José-María Sánchez-González [8,*], Esteban Moreno-Toral [9], Ana-Luisa González [10], Mike Zein [11], Pablo Larco, Jr. [11], Carolina Mercado [11] and Maria-Adelaida Piedrahita [12]

1 Department of Ophthalmology, Instituto de la Visión, Quevedo 660003, Ecuador
2 Department of Ophthalmology, Research and Ocular Health Group, Unbosque, University of El Bosque, Bogota 110111, Colombia
3 Department of Ophthalmology, Qvision, VITHAS Almería Hospital, 04120 Almeria, Spain
4 Department of Ophthalmology, Regional Universitary Hospital of Málaga, 18014 Granada, Spain
5 Department of Surgery, Ophthalmology Area, University of Seville, 41012 Seville, Spain
6 Department of Opthalmology, Vithas Malaga, 29016 Malaga, Spain
7 Department of Ophthalmology, National Eye Institute, National Institute of Health, Bethesda, MD 20892, USA
8 Department of Physics of Condensed Matter, Optics Area, University of Seville, 41012 Seville, Spain
9 Department of Pharmacy and Pharmaceutical Technology, University of Seville, 41012 Seville, Spain
10 Department of Ophthalmology, Research Department Clínica La Luz, Lima 15046, Peru
11 Department of Ophthalmology, Bascom Palmer Eye Institute, School of Medicine, University of Miami Miller, Miami, FL 33136, USA
12 Department of Ophthalmology, School of Medicine, University CES, Medellin 050021, Colombia
* Correspondence: jsanchez80@us.es

Abstract: Dry eye has two basic subdivisions: aqueous deficient dry eye (ADDE), with SS a major cause; and evaporative dry eye (EDE), due to either intrinsic or extrinsic factors. SS is a chronic inflammatory disorder defined by dysfunction of the exocrine glands leading to dry eye and dry mouth. The objective of this article was to carry out a systematic and critical review of several scientific publications on dry eye disease, with the aim of providing general recommendations to distinguish dry eye and its different variants in patients with SS, during the period 1979 to 2020, using search engines for articles indexed in Scopus, Latindex, Scielo, Clinical Trials, Medline, Embase, and Cochrane, allowing the analysis of 132 articles published in indexed journals on the subject of dry eye disease and SS, evidencing its conceptualization, prevalence, risk factors, etiopathogenesis, clinical manifestations, diagnosis, and treatment.

Keywords: dry eye; Sjögren's syndrome; evaporative dry eye; water deficiency dry eye; questionnaires; Schirmer I test; Schirmer II test; invasive tear film rupture time; tear meniscus height

Citation: Barrientos, R.T.; Godín, F.; Rocha-De-Lossada, C.; Soifer, M.; Sánchez-González, J.-M.; Moreno-Toral, E.; González, A.-L.; Zein, M.; Larco, P., Jr.; Mercado, C.; et al. Ophthalmological Approach for the Diagnosis of Dry Eye Disease in Patients with Sjögren's Syndrome. *Life* **2022**, *12*, 1899. https://doi.org/10.3390/life12111899

Academic Editor: Akio Oishi

Received: 21 October 2022
Accepted: 14 November 2022
Published: 15 November 2022

1. Introduction

Sjögren's syndrome (SS) is a chronic, systemic autoimmune disease that causes the dysfunction of the exocrine glands. It includes immune-mediated damage to the lacrimal and salivary glands, destroyed by infiltrating lymphocytes [1,2]. Dry eye disease (DED) is a multifactorial disease of the ocular surface characterized by a loss of tear film homeostasis and accompanied by ocular symptoms in which the instability and hyperosmolarity of the tear film, inflammation and ocular surface damage, and neurosensory abnormalities play an etiological role [3–6]. Symptoms can vary from itching or sandy to burning and stinging sensation. Diagnosis begins with determining the dry eye essential nature: aqueous deficient dry eye (ADDE), or evaporative dry eye (EDE) [1]. SS depends on the ADDE division, furthermore ADDE cases need to be investigated as potential SS associated dry eye. In Figure 1, the leading etiological causes of DED are represented [3–6].

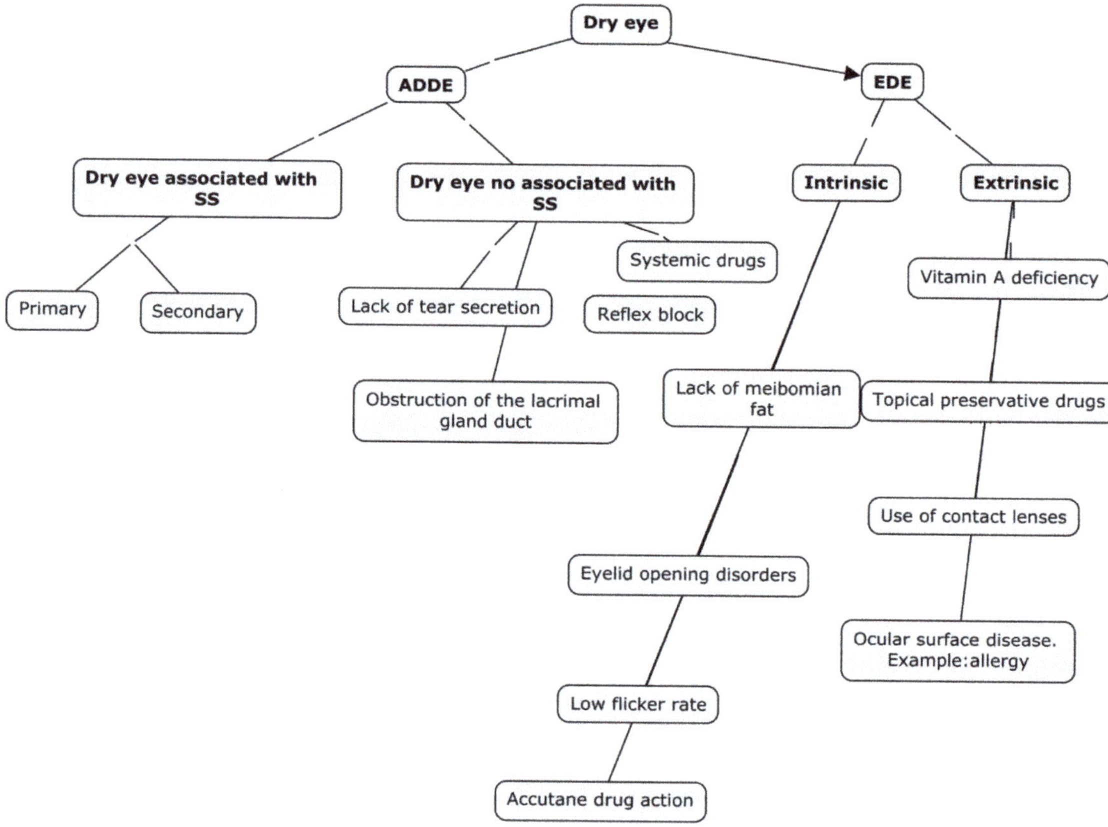

Figure 1. Main etiological causes of dry eye. Dry eye due to lack of aqueous secretion has two main groups: dry eye associated with SS and dry eye not associated with SS. ADDE: Aqueous deficient dry eye. EDE: Evaporative dry eye. SS: Sjögren Syndrome.

Since the 2017 TFOS DEWS II report, additional risk factors for SS have been identified, associating it with age, female gender, poor health, use of contact lenses, smoking, use of oral steroids or antidepressants, poorly managed thyroid disease and a greater extent of medical comorbidities, occupational risk factors (prolonged screen time), and environment factors (air conditioning and heaters) [5,7]. Factors that lowered the risk include a sedentary lifestyle and the use of angiotensin-converting enzyme inhibitors [8,9].

The human eye is usually protected from evaporation and desiccation by tear film homeostasis, which regulates the secretion of tears and distribution in the ocular surface in response to the blinking reflex. DED is characterized by a low quantity or quality of tears, destabilizing this microenvironment. The principal mechanism in DED is evaporative loss of water that leads to tearing hyperosmolarity. These mechanisms are thought to drive inflammation of the ocular surface and cell apoptosis in both the epithelial cells of the cornea and conjunctiva and the goblet cells of the conjunctiva [10]. In the case of SS, the etiology is not entirely clear. The presence of salivary gland epithelial cells expressing primary histocompatibility complex class II molecules and the identification of specific markers such as HLA-DR15 and HLA-DR3 imply that there are environmental antigens that trigger an inflammatory response [11]. The following is the recommended sequence (Table 1) of the diagnostic workup for DED:

Table 1. Sequence of diagnostic tests for DED.

1.	History
2.	Use of symptom questionnaires.
3.	Determine functional VA and contrast sensitivity
4.	Measure the invasive tear film break-up time (BUT)
5.	Ocular surface staining with fluorescein sodium
6.	Ocular surface staining with lissamine green
7.	Evaluation of tear flow using the Schirmer I–II test
8.	Perform an ocular surface examination
9.	Meibometrics
10.	Measurement of tear osmolarity

BUT: invasive tear film breakup time; Schirmer I test; Schirmer test. (Adapted from Merayo L, J. Spanish guidelines for the treatment of eye disease. Consensus document. Spanish Society of Ocular Surface and Cornea. 2017; 4 (1): 33).

1.1. Anamnesis

The patient should be asked about symptoms suggestive of dry eye. These include eye irritation, burning, stinging, foreign body sensation, blurred vision, improved vision when blinking, photophobia, or pain. The patient should also be asked about possible risk factors such as a history of collagen diseases, refractive surgery, stem cell transplantation, hepatitis, vitamin A deficiency, antihistamines, selective serotonin reuptake inhibitors, tricyclic antidepressants, and beta-blockers [9,12]. If the patient already has a diagnosis of SS, it is crucial to learn if this diagnosis was confirmed through gland biopsy or laboratory markers, Table 2.

Table 2. Directed interrogation of ocular and medical history.

Ocular
• Use of topical treatments: frequency, duration, effects, and whether or not they contain preservatives (artificial tears, glaucoma treatments, corticosteroids, antihistamines, vasoconstrictors and phytotherapy preparations, etc.). • Use of contact lenses: frequency, and care. • Allergic conjunctivitis. • History of eye or eyelid surgery. • History of ocular surface diseases. • Facial paralysis.
Medical history
• Smoking (including passive smoking). • Facial and eyelid hygiene (products, techniques, and frequency). • Dry mouth sensation. • Fatigue. • Joint and muscle pain. • Use of systemic drugs (diuretics, antihistamines, hormonal treatments, antidepressants, antineoplastics and any anticholinergic drug, etc.). • History of systemic inflammatory diseases (Sjögren's syndrome, rheumatoid arthritis and systemic lupus erythematosus, etc.). • Menopause. • Trauma (mechanical, thermal, and chemical). • Atopy. • Head and neck surgeries or transplants. • Chronic viral infections. • Neurological disorders (Parkinson's disease and trigeminal neuralgia, among others).

(Adapted from Merayo L, J. Spanish guidelines for the treatment of eye disease. Consensus document. Spanish Society of Ocular Surface and Cornea. 2017; 4 (1): 34).

1.2. Questionnaires on Symptoms

Numerous questionnaires are available for the assessment of symptoms in patients with DED, the TFOS DEWS II report recommends OSDI and DEQ-5 questionnaires. These questionnaires are both completed by the patient [7,9].

1.3. Determination of the Functional Visual Acuity (FVA)

The FVA is the continuous VA of the patient when performing activities such as reading or driving. However, standard VA tests may be normal [13].

1.4. Measurement of the Invasive Tear Film Breakdown Time (BUT)

The measurement of the invasive tear film breakdown time (BUT) is defined as the time between a complete blink and the appearance of the first tear film rupture. The average value is 10 s or more [14–16].

1.5. Lacrimal Osmolarity

Tear osmolarity was performed using TearLab® (TearLab Corporation, San Diego, CA, USA). The cutoff suggestion ranges were 300–320 mOsm/L in mild DED, and 320–340 mOsm/L in moderate DED [17–20]. This test must be taken before the instillation of any drops or dyes on the ocular surface so it is the first thing that should be completed, with a differential value between eyes of > 8 mOsm/L) as a reference of positivity to specify dry eye [21–23].

It should be noted that the frequency of meibomian gland dysfunction (MGD) is higher in patients with SS than in the average population, which contributes to the worsening clinical picture. The complete diagnostic analysis for SS is used based on the European–American criteria (ACR/EULAR) proposed in 2016 for the classification of primary SS (Table 3) [24–26]. The more recent criteria are considered more refined and emphasize objective measures, including biopsy samples, as opposed to older criteria that were more exhaustive [25]. The biopsy sample study improves SS diagnosis but needs an interdisciplinary consultant with a rheumatologist [27–29]. When SS is suspected in a patient with dry eye, knowledge of the extraocular signs and symptoms of SS can aid in making the diagnosis [30–32].

Table 3. European–American criteria (ACR/EULAR) proposed in 2016 for the classification of primary SS.

Item	Score
Focal lymphocytic sialadenitis in minor salivary gland with $\geq$ 1 lymphocytic focus/4 mm^2 of glandular tissue	3
Anti-SSA/Ro positive	3
Ocular staining score $\geq$ 5 (or $\geq$ 4 according to the Bjsterveld scale) in at least one eye	1
Schirmer test $\leq$ 5 mm/5 min, in at least one eye	1
Unstimulated salivary flow $\leq$ 0.1 mL/minute	1
Diagnosis > 4 points	

(Adapted from Shiboski CH, et al. Ann Rheum Dis 2017; 76: 9–16. doi:10.1136/annrheumdis-2016-210571).

1.6. Measurement of Matrix Metalloproteinases (MMP-9)

In recent years the technology has allowed clinical ophthalmologists, and other health-care staff to measure MMP-9 levels in the tear film with a simple device, the InflammaDry assay (Rapid Pathogen Screening, Inc., Sarasota, FL, USA) [33,34]. The test is a rapid (<10 min) and noninvasive test that measures MMP-9 levels above 40 ng/mL. The results depend on two lines, one blue and one pink. Only the blue line on the test windows indicates a level under 40 ng/mL and two lines indicate levels above this mark [35].

1.7. Measurement of the Ocular Surface Sodium Fluorescein Staining

Following Sjögren's international collaborative clinical alliance (SICCA) registry ocular examination protocol, the cornea staining score was determined [36]. Fluorescein stains areas of the ocular surface with corneal epithelial defects; its application allows for a better appreciation of the size of the tear meniscus. This technique should be performed before positioning the patient for evaluation and without lighting to avoid the lacrimation reflex and the formation of mucous filaments [37].

1.8. Dye Staining (Lissamine Green and Rose Bengal) of the Ocular Surface

Damage to the ocular surface can be examined with specific stains where defects of the corneal epithelium (rose bengal) and cell damage of the cornea and conjunctiva (lissamine green) can be seen and both must be performed separately [38]. In Table 4 the characteristics of the different colorants are compared [39].

Table 4. Comparison of the dyes fluorescein, rose bengal and lissamine green.

	Fluorescein	Bengal Rose	Lissamine Green
Healthy cell staining	No	Yes	No
Dead cell staining	No	Yes	Yes
Stain significance	Cell disruption and increase membrane permeability	Loss or insufficient protection of ocular surface mucin	Cell degeneration and death
Staining best seen with	Yellow barrier filter	Green barrier filter	Red barrier filter

(Adapted from Tseng SCG. Evaluation of the ocular surface in dry-eye conditions. Int Clin Ophthalmol 1994; 34: 57–69).

The classification system first proposed by van Bijsterveld, on which current diagnostic criteria are based, qualifies three areas in each eye: the nasal and temporal bulbar conjunctiva and the cornea. The intensity of rose bengal staining is rated on a scale of 0 (no staining) to 3 (confluent staining) for each area (Figure 2). The maximum staining value for each eye is nine. Staining values of three or more are considered abnormal [8,40,41].

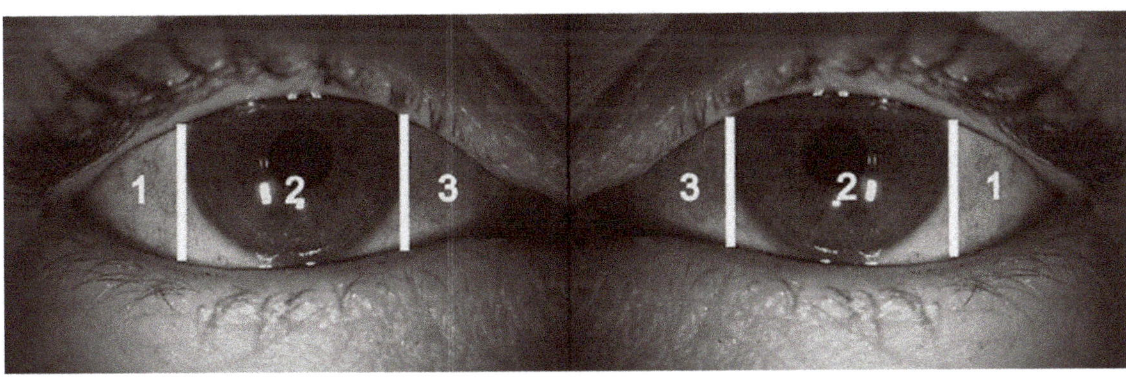

Figure 2. The exposed interpalpebral portions of the nasal and temporal conjunctiva and cornea are graded on a scale of 0 (no staining) to 3 (confluent staining). The maximum possible total score for each eye is nine. A score greater than three is considered abnormal.

1.9. Tear Flow Evaluation Using the Schirmer I–II Test

The Schirmer test measures the basal and reflex tear secretion from the primary and accessory lacrimal glands and the volume of the marginal tear film. It is performed by placing a strip of special millimeter paper between the outer half of the lower eyelid and the bulbar conjunctiva of each eye, and then the patient keeps his eyes closed for 5 min. Then the strips of paper are removed, and it is observed how wet the paper is. Various cut-off values have been proposed for the diagnosis of DED, between $\leq$5 mm and $\leq$10 mm [7,42]. The test can be performed under topical anesthesia or without anesthesia.

1.10. Perform an Examination of the Ocular Surface

It is advisable to thoroughly inspect the external surface and evaluate the ease of expression of the meibomian glands and permeability of the glandular orifices in search of MGD. Patients with mild SS can present with a normal eye examination and preserved tear function. Severe SS can present with cicatricial conjunctivitis leading to fibrosis and scarring [1,3,5,6].

1.11. Evaluation of the Meibomian Glands (Meibometry)

The meibomian glands should be assessed for the volume, quality, and ease of expression of secretions. To assess the quality of secretion, it is measured with meibometry where each gland in the central zone of the lower lid is gently pressed, and the secretion of each gland is scored as 0 (clear or normal), 1 (cloudy), 2 (granular), and 3 (pasty, similar to toothpaste) [43]. The meiboscore is the assessment of the expressibility of the glands, it is studied by pressing five glands of the lower or upper eyelid with a cotton swab [44]. The score for the number of squeezable glands is 0 (all glands), 1 (three or four glands), 2 (one or two glands), and 3 (no glands) [43,45,46].

2. Treatment of DED

TFOS DEWS II recommends individualized management of DED due to lack of aqueous and evaporative secretion, as well as the overall severity of the disease [3,5]. A dry eye can indicate the presence of SS, particularly when it is associated with inflammation, difficulty in medical management, or dry mouth. A patient with suspected SS should have an interdisciplinary assessment and follow up. The first step to diagnosing and managing this disease is referring to a rheumatologist for systemic treatment. The patient should also be referred to a dentist for the prevention and management of oral diseases [47]. Treatment for DED progresses gradually, beginning with education, diet modification, eyelid hygiene, lubricating eye drops, environmental factors modification, and nonpharmacological, and pharmacological management (Table 5) [48–50].

Diet modifications to treat DED [51,52] include supplementation with essential fatty acids (i.e., omega-3, omega-6/gamma-linolenic acid, or both) [25]. In addition, increasing fluid intake to ensure adequate general hydration [5] and avoiding alcohol intake have been recommended [53].

Table 5. Treatment schedule for DED.

Step	Treatment
Step 1	Education of the patient about: • Management • Possible dietary modifications • Treatment • Forecast • Modifications of the local environment. • Identification and modifications or elimination of systemic problems and topical medications • Administration of warm compresses • Perform eyelid hygiene and hot compresses • Administration of eye lubricants

Table 5. *Cont.*

Step	Treatment
Step 2	If the options in Step one are inadequate: • Manage ○ Unpreserved eye lubricants ○ Tea tree oil treatment for Demodex (if present) ○ Night treatments ○ Humidity chamber device ○ Ointments ○ Prescription drugs: doxycycline and tetracycline ○ Topical antibiotics or antibiotics and steroid combination ○ Topical corticosteroid ○ Nonglucocorticoid topical immunomodulatory-type drugs (such as cyclosporine) ○ Topical LFA-1 antagonist medication (such as Lifitegrast) ○ Oral macrolide or tetracycline antibiotics • Preserve tears with punctal occlusion or moisture, camera glasses • Complete therapies in the office ○ Intense Pulsed Light (IPL) therapy. Five sessions, with an interval of 3–4 weeks. Maintenance every 6–12 months. ○ Meibomian gland expression
Step 3	If the options in steps one and two are inadequate: • Administer oral secretagogues • Autologous or allogeneic serum eye drops • Use of therapeutic contact lenses • Soft bandage lenses • Rigid scleral lenses
Step 4	If the options in the previous steps are inadequate: • Administer topical corticosteroids for longer duration. • Perform surgical options • Amniotic membrane graft • Surgical punctal occlusion • Tarsorrhaphy • Salivary gland transplant

Abbreviations: DEWS II, Dry Eye Workshop II; LFA-1, lymphocyte function-associated antigen 1; IPL, intense pulsed light. (Adapted from Jones L, Downie LE, Korb D, et al. TFOS DEWS II management and therapy report. Ocul Surf. 2017; 15: 575–628.).

With regard to complementary medicine, there is some evidence from clinical trials supporting the use of traditional Chinese herbs and acupuncture [51–53]. Using over-the-counter hot compresses, artificial tears, or other eye lubricants can be used as first-line treatments [4]. Measures for eyelid hygiene include detergent-based cleaning products and microblepharon-exfoliation procedures to help remove residue from the eyelid margin. Hygienic saline containing 0.01% pure hypochlorous acid has been shown to reduce biofilm, and hot compresses for eyelid hyperthermia are commonly used to soften the meibum and facilitate its exit from the ducts. Alternative options include topical corticosteroids for a limited duration, topical cyclosporine 0.05%, tacrolimus 0.03%, and lifitegrast 5%. Antibiotics such as oral doxycycline can also be given for two to three months [51,54–56].

Devices that can be used to preserve or stimulate tears include silicone-based punctal occlusion in thermolabile polymer and hydrogel devices, therapeutic contact lenses, and intranasal tear stimulation (e.g., TrueTear®, Allergan, Pleasanton, CA, USA). MGD can be treated by meibomian gland expression and devices such as vectorized thermal pulsation therapy (i.e., LipiFlow®, Johnson & Johnson Vision, Jacksonville, FL, USA), intense pulsed light (IPL, i.e., Optima IPL M22, Lumenis, Salt Lake City, UT, USA), light-based heat and compression (i.e., iLux, Alcon, Fort Worth, TX, USA), and portable thermal energy therapy (i.e., TearCare, Sight Sciences, Menlo Park, CA, USA) [4,47,57–59]. The third care step includes oral secretagogues and autologous or allogeneic eye drops. In contrast, step four includes topical corticosteroids for longer durations, amniotic membrane grafts, surgical punctal occlusion, and more complex surgical approaches.

3. Discussion

There is a high tendency in recent times to diagnose dry eye disease worldwide due to the multifactorial nature of this entity [60]. However, the casuistry collected in much of the literature shows that Caucasian populations with a prevalence of 0.04%, individuals older than 65 years, and females tend to have a more significant association. This low-rate result was from using autoantibodies to classify patients and according to Hochberg, it was 0.6% in Greece [43]. Similar results were found in Slovenia (0.6%), Denmark (0.6–21.11%), and the United Kingdom (3–4%) using the European criteria, however in the latter using the American–European consensus, the prevalence ranged from 0.1 to 0.4% [14,58]. Using the Copenhagen criteria, the prevalence was 2.7% in Sweden and 0.7% in China [14,60,61]. In Latin America, in Brazil it was 0.17% [62], Argentina it was 0.17% [63], using the COPCORD methodology, and in Colombia it was 0.12%, with the American–European classification criteria employed in all of these [64].

There are few demographic and characterization studies of SS in Latin America, while, in other countries, important information on its management, diagnosis, and treatment is available. In a study carried out in Colombia in 2016, with 58,680 cases, they found a prevalence in people over 18 years of 0.12%. Eighty-two percent were women, with a 4.6:1 female:male ratio and there was a higher prevalence among the 65 to 69 age group. In Ecuador, Oviedo and Moya (2019), reported a prevalence of dry eye disease in a population that varied between 27 and 88% according to the OSDI, McMonnies, and DEQ5 questionnaire, estimating a range of 27–34.5% with a median age of 34 years [64–66]. These findings are accentuated by the lack of timely medical attention and failure to reach a timely diagnosis [50].

In SS, dry eye and mouth have been reported in up to 30% of people over 65 years of age, particularly in women in their perimenopausal and postmenopausal years [67]. Ocular signs include hyperemia, conjunctival keratinization, punctate or filamentous keratitis, and in some cases, involvement of the eyelids [68]. At the ocular level, the examinations focus on the objective evaluation of tear production, stability, osmolarity, and evaluation of the lid margin and the ocular surface [69]. In general, ocular treatment includes artificial tears as the first treatment alternative in order to increase the volume of the tear film and reduce friction, topical corticosteroids, immunomodulatory agents, immunosuppressants, autologous serum, and in experimental studies, new treatments with cells are proposed (mesenchymal stem cells or multipotent stem cells (MSC)) [68], in addition to systemic treatments to treat extraglandular manifestations.

The relationship between some signs and symptoms in patients with Sjögren's syndrome represents an example of "heterogeneity" associated with DED [26,70,71]. It is commonly associated with two possible etiologies: a water deficiency and excessive evaporation.

Although several studies have not reported a strong association or correlation between symptoms and signs in patients with dry eye assessed from the use of questionnaires and clinical tests [66] such as Begley et al. [45], Schein et al. [19], Hay et al. [63], Nichols et al. [32], and Lin et al. [72], most of them have found a weak or moderate association between these parameters, including a high correlation between the clinical diagnosis and the patient's symptoms, suggesting that the symptoms have a more significant influence on the diagnosis of dry eye than the results of clinical tests [53].

Many other studies have reported an association or correlation between signs and symptoms of irritation in patients with dry eye due to meibomian gland disease or aqueous lacrimal deficiency, using various clinical tests such as Schirmer, BUT, fluorescein staining, clearing test of fluorescein, and corneal sensation (Afonso et al. [31], 1999; Macri & Pflugfelder, Pflugfelder et al. [39]) [16,64,73,74]. In terms of tear osmolarity, although it shows a very close agreement between the eyes and in the same eyes over time in normal subjects, it shows increasing variability in subjects with dry eye (Sullivan et al. [75] 2014). This is believed to be due to the instability of the tear film in affected patients and can be used as a diagnostic hallmark of DED [32,63,76]. These results advocate the clinical utility of a consensus of signs, which better captures the entire disease and discourages dependence on symptoms alone [77]. This finding differs from that reported by Schein et al. [19], who found no association between the presence of more frequent symptoms and a lower Schirmer result, regardless of whether the analysis was based on mean scores with a cut-off value of five or a cut-off value of seven [77]. Their sensitivity and specificity in the detection of symptomatic subjects was low [72,78,79].

Questionnaires are of some value for the evaluation of the etiology of dry eye [78,79]. However, alone they are insufficient to confirm a diagnosis of SS since they have no cost if they have a high value of sensitivity and specificity [13,80]. Tear function tests such as tear meniscus height and BUT play a role in the differential diagnosis of SS, the most important aspect being the difference between ADDE and EDE accompanied by the performance of the Schirmer I–II tests, as discussed above. The tear function index has been reported to be helpful in the diagnosis of SS [81–84]. In addition, vital rose bengal or lissamine green staining of the interpalpebral fissure is a noninvasive way to help diagnose SS, as conjunctival staining may be seen earlier in the course of the disease [85,86].

Regarding future lines of research, the noninvasive diagnostic tools for SS diagnosis and DED examination should be implemented in all ophthalmologist and optometrist scientific communities because of the the reliability and repeatability of the measurements. This allows the measurement of noninvasive breakup time which is more unambiguous to interpret than the one measured with fluorescein (as the FBUT might be related with local thinning of the tear film rather than with actual breakup event) and has been shown as potentially having a stronger correlation with patient discomfort. Furthermore, noninvasive techniques can be readily used by a broader range of medical personnel, such as technicians and nurses, which allows for more rapid and broadly available diagnostics of DED [87–90].

Another interesting future research line is the evaluation of the ocular surface microbiota in the pathogenesis and management of different eye diseases. Recently a larger multicenter study proposed the concept of eye community state type (ECST) with the aim of stratifying the different profiles of bacterial communities that coexist in a healthy eye. It was observed that nine different ECST could be considered within the healthy bacterial population [91]. However, the central job of the ocular surface and oral microbiota in the pathogenesis of SS is not completely understood, although microbiota changes have been distinguished [92] in these patients. Bacterial mimicry has been proposed as one of the systems by which the microbiome may take part in illness acceptance. Furthermore, microbiota dysbiosis in SS suggest that lower diversity may lead to higher disease activity [92]. Other studies have found different outcomes with greater phylogenetic diversity [93]. Therefore, commensal microbes could play a fundamental part in the pathogenesis of SS [94]. Peptides obtained from oral, stomach, and skin commensal microscopic organisms might prompt an insusceptible reaction by initiating the Ro60-receptive immune system microorganisms [94]. Likewise, it seems that an alteration of commensal bacteria in the gut caused a worsening of dry eye in SS. An improvement in microbiome health could improve the condition [93,95]. Nonetheless, the exact role of the role of the microbiota both in the management and in the diagnosis of this pathology should continue to be studied in multicenter studies with a larger number of patients.

4. Conclusions

Patient grumblings and clinical discoveries that are reminiscent of dry eye, particularly with ADDE, should always be considered possible indications of SS and a they should be given a brief further examination. Given the accessibility of new serologic indicative tests and the possibly extreme results of deferring a determination, the examination should include requests about corresponding side effects of oral dryness and a serologic assessment. Dry eye related with SS is not restricted to ADDE; associative MGD and EDE are frequently seen.

The ocular manifestations of SS are often accompanied by oral or systemic manifestations and a certain humeral profile. A severe dry eye accompanied by these systemic manifestations clarifies the diagnosis and directs a change in treatment from systemic treatment to a sequential ocular treatment.

There is no curative treatment available for SS, so a comprehensive treatment of the patient is essential: education and information, vigilance, and proactive steps by ophthalmologists and optometrists, in conjunction with rheumatology specialists, play a fundamental role and facilitate the early recognition of SS, allowing the management and timely intervention of ocular and systemic manifestations.

Author Contributions: Conceptualization, R.T.B., F.G., C.R.-D.-L., M.S., J.-M.S.-G., E.M.-T., A.-L.G., M.Z., P.L.J., C.M. and M.-A.P.; methodology, R.T.B., F.G., C.R.-D.-L., M.S., J.-M.S.-G., E.M.-T., A.-L.G., M.Z., P.L.J., C.M. and M.-A.P.; software, R.T.B., F.G., C.R.-D.-L., M.S., J.-M.S.-G., E.M.-T., A.-L.G., M.Z., P.L.J., C.M. and M.-A.P.; validation, R.T.B., F.G., C.R.-D.-L., M.S., J.-M.S.-G., E.M.-T., A.-L.G., M.Z., P.L.J., C.M. and M.-A.P.; formal analysis, R.T.B., F.G., C.R.-D.-L., M.S., J.-M.S.-G., E.M.-T., A.-L.G., M.Z., P.L.J., C.M. and M.-A.P.; investigation, R.T.B., F.G., C.R.-D.-L., M.S., J.-M.S.-G., E.M.-T., A.-L.G., M.Z., P.L.J., C.M. and M.-A.P.; resources, R.T.B., F.G., C.R.-D.-L., M.S., J.-M.S.-G., E.M.-T., A.-L.G., M.Z., P.L.J., C.M. and M.-A.P.; data curation, R.T.B., F.G., C.R.-D.-L., M.S., J.-M.S.-G., E.M.-T., A.-L.G., M.Z., P.L.J., C.M. and M.-A.P.; writing—original draft preparation, R.T.B., F.G., C.R.-D.-L., M.S., J.-M.S.-G., E.M.-T., A.-L.G., M.Z., P.L.J., C.M. and M.-A.P.; writing—review and editing, R.T.B., F.G., C.R.-D.-L., M.S., J.-M.S.-G., E.M.-T., A.-L.G., M.Z., P.L.J., C.M. and M.-A.P.; visualization, R.T.B., F.G., C.R.-D.-L., M.S., J.-M.S.-G., E.M.-T., A.-L.G., M.Z., P.L.J., C.M. and M.-A.P.; supervision, R.T.B., F.G., C.R.-D.-L., M.S., J.-M.S.-G., E.M.-T., A.-L.G., M.Z., P.L.J., C.M. and M.-A.P.; project administration, R.T.B., F.G., C.R.-D.-L., M.S., J.-M.S.-G., E.M.-T., A.-L.G., M.Z., P.L.J., C.M. and M.-A.P.; funding acquisition, R.T.B., F.G., C.R.-D.-L., M.S., J.-M.S.-G., E.M.-T., A.-L.G., M.Z., P.L.J., C.M. and M.-A.P. All authors have read and agreed to the published version of the manuscript.

Funding: This research received no external funding.

Institutional Review Board Statement: Not applicable.

Informed Consent Statement: Not applicable.

Data Availability Statement: Not applicable.

Conflicts of Interest: The authors declare no conflict of interest.

References

1. The Definition and Classification of Dry Eye Disease: Report of the Definition and Classification Subcommittee of the International Dry Eye Workshop (2007)—ScienceDirect. Available online: https://www.sciencedirect.com/science/article/abs/pii/S1542012412700812?via%3Dihub (accessed on 17 October 2022).
2. Tincani, A.; Andreoli, L.; Cavazzana, I.; Doria, A.; Favero, M.; Fenini, M.-G.; Franceschini, F.; Lojacono, A.; Nascimbeni, G.; Santoro, A.; et al. Novel Aspects of Sjögren's Syndrome in 2012. *BMC Med.* **2013**, *11*, 93. [CrossRef] [PubMed]
3. Stapleton, F.; Alves, M.; Bunya, V.Y.; Jalbert, I.; Lekhanont, K.; Malet, F.; Na, K.-S.; Schaumberg, D.; Uchino, M.; Vehof, J.; et al. TFOS DEWS II Epidemiology Report. *Ocul. Surf.* **2017**, *15*, 334–365. [CrossRef] [PubMed]
4. Jones, L.; Downie, L.E.; Korb, D.; Benitez-Del-Castillo, J.M.; Dana, R.; Deng, S.X.; Dong, P.N.; Geerling, G.; Hida, R.Y.; Liu, Y.; et al. TFOS DEWS II Management and Therapy Report. *Ocul. Surf.* **2017**, *15*, 575–628. [CrossRef] [PubMed]
5. Craig, J.P.; Nelson, J.D.; Azar, D.T.; Belmonte, C.; Bron, A.J.; Chauhan, S.K.; de Paiva, C.S.; Gomes, J.A.P.; Hammitt, K.M.; Jones, L.; et al. TFOS DEWS II Report Executive Summary. *Ocul. Surf.* **2017**, *15*, 802–812. [CrossRef] [PubMed]
6. Willcox, M.D.P.; Argüeso, P.; Georgiev, G.A.; Holopainen, J.M.; Laurie, G.W.; Millar, T.J.; Papas, E.B.; Rolland, J.P.; Schmidt, T.A.; Stahl, U.; et al. TFOS DEWS II Tear Film Report. *Ocul. Surf.* **2017**, *15*, 366–403. [CrossRef] [PubMed]
7. Brito-Zerón, P.; Retamozo, S.; Kostov, B.; Baldini, C.; Bootsma, H.; De Vita, S.; Dörner, T.; Gottenberg, J.-E.; Kruize, A.A.; Mandl, T.; et al. Efficacy and Safety of Topical and Systemic Medications: A Systematic Literature Review Informing the EULAR Recommendations for the Management of Sjögren's Syndrome. *RMD Open* **2019**, *5*, e001064. [CrossRef]
8. Schaumberg, D.A.; Sullivan, D.A.; Dana, M.R. Epidemiology of Dry Eye Syndrome. *Adv. Exp. Med. Biol.* **2002**, *506*, 989–998. [CrossRef]
9. Oden, N.L.; Lilienfeld, D.E.; Lemp, M.A.; Nelson, J.D.; Ederer, F. Sensitivity and Specificity of a Screening Questionnaire for Dry Eye. *Adv. Exp. Med. Biol.* **1998**, *438*, 807–820. [CrossRef]
10. Chia, E.-M.; Mitchell, P.; Rochtchina, E.; Lee, A.J.; Maroun, R.; Wang, J.J. Prevalence and Associations of Dry Eye Syndrome in an Older Population: The Blue Mountains Eye Study. *Clin. Exp. Ophthalmol.* **2003**, *31*, 229–232. [CrossRef]

11. Gottenberg, J.-E.; Busson, M.; Loiseau, P.; Cohen-Solal, J.; Lepage, V.; Charron, D.; Sibilia, J.; Mariette, X. In Primary Sjögren's Syndrome, HLA Class II Is Associated Exclusively with Autoantibody Production and Spreading of the Autoimmune Response. *Arthritis Rheum.* **2003**, *48*, 2240–2245. [CrossRef]
12. Rieger, G. The Importance of the Precorneal Tear Film for the Quality of Optical Imaging. *Br. J. Ophthalmol.* **1992**, *76*, 157–158. [CrossRef]
13. Begley, C.G.; Chalmers, R.L.; Abetz, L.; Venkataraman, K.; Mertzanis, P.; Caffery, B.A.; Snyder, C.; Edrington, T.; Nelson, D.; Simpson, T. The Relationship between Habitual Patient-Reported Symptoms and Clinical Signs among Patients with Dry Eye of Varying Severity. *Investig. Ophthalmol. Vis. Sci.* **2003**, *44*, 4753–4761. [CrossRef] [PubMed]
14. Liew, M.S.H.; Zhang, M.; Kim, E.; Akpek, E.K. Prevalence and Predictors of Sjogren's Syndrome in a Prospective Cohort of Patients with Aqueous-Deficient Dry Eye. *Br. J. Ophthalmol.* **2012**, *96*, 1498–1503. [CrossRef] [PubMed]
15. van Bijsterveld, O.P. Diagnostic Tests in the Sicca Syndrome. *Arch. Ophthalmol.* **1969**, *82*, 10–14. [CrossRef] [PubMed]
16. Lemp, M.A.; Bron, A.J.; Baudouin, C.; Benítez Del Castillo, J.M.; Geffen, D.; Tauber, J.; Foulks, G.N.; Pepose, J.S.; Sullivan, B.D. Tear Osmolarity in the Diagnosis and Management of Dry Eye Disease. *Am. J. Ophthalmol.* **2011**, *151*, 792–798.e1. [CrossRef]
17. Tashbayev, B.; Utheim, T.P.; Utheim, Ø.A.; Ræder, S.; Jensen, J.L.; Yazdani, M.; Lagali, N.; Vitelli, V.; Dartt, D.A.; Chen, X. Utility of Tear Osmolarity Measurement in Diagnosis of Dry Eye Disease. *Sci. Rep.* **2020**, *10*, 5542. [CrossRef]
18. Onuora, S. First EULAR Recommendations for Sjögren Syndrome Published. *Nat. Rev. Rheumatol.* **2020**, *16*, 2. [CrossRef]
19. Schein, O.D.; Muñoz, B.; Tielsch, J.M.; Bandeen-Roche, K.; West, S. Prevalence of Dry Eye among the Elderly. *Am. J. Ophthalmol.* **1997**, *124*, 723–728. [CrossRef]
20. Lemp, M.A. Report of the National Eye Institute/Industry Workshop on Clinical Trials in Dry Eyes. *CLAO J.* **1995**, *21*, 221–232.
21. Ohashi, Y.; Ishida, R.; Kojima, T.; Goto, E.; Matsumoto, Y.; Watanabe, K.; Ishida, N.; Nakata, K.; Takeuchi, T.; Tsubota, K. Abnormal Protein Profiles in Tears with Dry Eye Syndrome. *Am. J. Ophthalmol.* **2003**, *136*, 291–299. [CrossRef]
22. Khanal, S.; Tomlinson, A.; Diaper, C.J.M. Tear Physiology of Aqueous Deficiency and Evaporative Dry Eye. *Optom. Vis. Sci. Off. Publ. Am. Acad. Optom.* **2009**, *86*, 1235–1240. [CrossRef] [PubMed]
23. Beckman, K.A. Detection of Early Markers for Sjögren Syndrome in Dry Eye Patients. *Cornea* **2014**, *33*, 1262–1264. [CrossRef] [PubMed]
24. Machetta, F.; Fea, A.M.; Actis, A.G.; de Sanctis, U.; Dalmasso, P.; Grignolo, F.M. In Vivo Confocal Microscopic Evaluation of Corneal Langerhans Cells in Dry Eye Patients. *Open Ophthalmol. J.* **2014**, *8*, 51–59. [CrossRef] [PubMed]
25. Shiboski, S.C.; Shiboski, C.H.; Criswell, L.A.; Baer, A.N.; Challacombe, S.; Lanfranchi, H.; Schiødt, M.; Umehara, H.; Vivino, F.; Zhao, Y.; et al. American College of Rheumatology Classification Criteria for Sjögren's Syndrome: A Data-Driven, Expert Consensus Approach in the Sjögren's International Collaborative Clinical Alliance Cohort. *Arthritis Care Res.* **2012**, *64*, 475–487. [CrossRef]
26. Versura, P.; Frigato, M.; Cellini, M.; Mulè, R.; Malavolta, N.; Campos, E.C. Diagnostic Performance of Tear Function Tests in Sjogren's Syndrome Patients. *Eye Lond. Engl.* **2007**, *21*, 229–237. [CrossRef]
27. Kaye, S.B.; Sims, G.; Willoughby, C.; Field, A.E.; Longman, L.; Brown, M.C. Modification of the Tear Function Index and Its Use in the Diagnosis of Sjögren's Syndrome. *Br. J. Ophthalmol.* **2001**, *85*, 193–199. [CrossRef]
28. Xuan, J.; Shen, L.; Malyavantham, K.; Pankewycz, O.; Ambrus, J.L.; Suresh, L. Temporal Histological Changes in Lacrimal and Major Salivary Glands in Mouse Models of Sjogren's Syndrome. *BMC Oral Health* **2013**, *13*, 51. [CrossRef]
29. Szalai, E.; Berta, A.; Szekanecz, Z.; Szûcs, G.; Módis, L. Evaluation of Tear Osmolarity in Non-Sjögren and Sjögren Syndrome Dry Eye Patients with the TearLab System. *Cornea* **2012**, *31*, 867–871. [CrossRef]
30. Williamson, J.F.; Huynh, K.; Weaver, M.A.; Davis, R.M. Perceptions of Dry Eye Disease Management in Current Clinical Practice. *Eye Contact Lens* **2014**, *40*, 111–115. [CrossRef]
31. Afonso, A.A.; Monroy, D.; Stern, M.E.; Feuer, W.J.; Tseng, S.C.; Pflugfelder, S.C. Correlation of Tear Fluorescein Clearance and Schirmer Test Scores with Ocular Irritation Symptoms. *Ophthalmology* **1999**, *106*, 803–810. [CrossRef]
32. Nichols, K.K.; Nichols, J.J.; Mitchell, G.L. The Lack of Association between Signs and Symptoms in Patients with Dry Eye Disease. *Cornea* **2004**, *23*, 762–770. [CrossRef] [PubMed]
33. Kook, K.Y.; Jin, R.; Li, L.; Yoon, H.J.; Yoon, K.C. Tear Osmolarity and Matrix Metallopeptidase-9 in Dry Eye Associated with Sjögren's Syndrome. *Korean J. Ophthalmol.* **2020**, *34*, 179–186. [CrossRef] [PubMed]
34. Messmer, E.M.; von Lindenfels, V.; Garbe, A.; Kampik, A. Matrix Metalloproteinase 9 Testing in Dry Eye Disease Using a Commercially Available Point-of-Care Immunoassay. *Ophthalmology* **2016**, *123*, 2300–2308. [CrossRef] [PubMed]
35. Park, J.Y.; Kim, B.G.; Kim, J.S.; Hwang, J.H. Matrix Metalloproteinase 9 Point-of-Care Immunoassay Result Predicts Response to Topical Cyclosporine Treatment in Dry Eye Disease. *Transl. Vis. Sci. Technol.* **2018**, *7*, 31. [CrossRef] [PubMed]
36. Whitcher, J.P.; Shiboski, C.H.; Shiboski, S.C.; Heidenreich, A.M.; Kitagawa, K.; Zhang, S.; Hamann, S.; Larkin, G.; McNamara, N.A.; Greenspan, J.S.; et al. A Simplified Quantitative Method for Assessing Keratoconjunctivitis Sicca from the Sjögren's Syndrome International Registry. *Am. J. Ophthalmol.* **2010**, *149*, 405–415. [CrossRef] [PubMed]
37. Meiners, P.M.; Vissink, A.; Kroese, F.G.M.; Spijkervet, F.K.L.; Smitt-Kamminga, N.S.; Abdulahad, W.H.; Bulthuis-Kuiper, J.; Brouwer, E.; Arends, S.; Bootsma, H. Abatacept Treatment Reduces Disease Activity in Early Primary Sjögren's Syndrome (Open-Label Proof of Concept ASAP Study). *Ann. Rheum. Dis.* **2014**, *73*, 1393–1396. [CrossRef]
38. Nichols, K.K.; Begley, C.G.; Caffery, B.; Jones, L.A. Symptoms of Ocular Irritation in Patients Diagnosed with Dry Eye. *Optom. Vis. Sci. Off. Publ. Am. Acad. Optom.* **1999**, *76*, 838–844. [CrossRef]
39. Pflugfelder, S.C.; Tseng, S.C.; Sanabria, O.; Kell, H.; Garcia, C.G.; Felix, C.; Feuer, W.; Reis, B.L. Evaluation of Subjective Assessments and Objective Diagnostic Tests for Diagnosing Tear-Film Disorders Known to Cause Ocular Irritation. *Cornea* **1998**, *17*, 38–56. [CrossRef]

40. Ramos-Casals, M.; Brito-Zerón, P.; Bombardieri, S.; Bootsma, H.; De Vita, S.; Dörner, T.; Fisher, B.A.; Gottenberg, J.-E.; Hernandez-Molina, G.; Kocher, A.; et al. EULAR Recommendations for the Management of Sjögren's Syndrome with Topical and Systemic Therapies. *Ann. Rheum. Dis.* **2020**, *79*, 3–18. [CrossRef]
41. Posso-Osorio, I.; Méndez-Rayo, T.; Soto, D.; Nieto-Aristizábal, I.; Cañas, C.A.; Tobón, G.J. Clinimetrics in Sjögren's Syndrome *Rev. Colomb. Reumatol. Engl. Ed.* **2019**, *26*, 260–267. [CrossRef]
42. Foulks, G.N.; Forstot, S.L.; Donshik, P.C.; Forstot, J.Z.; Goldstein, M.H.; Lemp, M.A.; Nelson, J.D.; Nichols, K.K.; Pflugfelder, S.C.; Tanzer, J.M.; et al. Clinical Guidelines for Management of Dry Eye Associated with Sjögren Disease. *Ocul. Surf.* **2015**, *13*, 118–132. [CrossRef] [PubMed]
43. Nichols, K.K.; Mitchell, G.L.; Zadnik, K. The Repeatability of Clinical Measurements of Dry Eye. *Cornea* **2004**, *23*, 272–285. [CrossRef] [PubMed]
44. Akpek, E.K.; Bunya, V.Y.; Saldanha, I.J. Sjögren's Syndrome: More Than Just Dry Eye. Available online: https://pubmed.ncbi.nlm.nih.gov/30681523/ (accessed on 17 October 2022).
45. Begley, C.G.; Chalmers, R.L.; Mitchell, G.L.; Nichols, K.K.; Caffery, B.; Simpson, T.; DuToit, R.; Portello, J.; Davis, L. Characterization of Ocular Surface Symptoms from Optometric Practices in North America. *Cornea* **2001**, *20*, 610–618. [CrossRef] [PubMed]
46. Methodologies to Diagnose and Monitor Dry Eye Disease: Report of the Diagnostic Methodology Subcommittee of the International Dry Eye WorkShop (2007). *Ocul. Surf.* **2007**, *5*, 108–152. [CrossRef]
47. Tsubota, K.; Hata, S.; Okusawa, Y.; Egami, F.; Ohtsuki, T.; Nakamori, K. Quantitative Videographic Analysis of Blinking in Normal Subjects and Patients with Dry Eye. *Arch. Ophthalmol.* **1996**, *114*, 715–720. [CrossRef] [PubMed]
48. Yoon, H.J.; Choi, W.; Yang, J.M.; Ji, Y.S.; Lee, S.-S.; Yoon, K.C. Characteristics of Dry Eye in Patients with Pre-Existing Sjögren's Syndrome According to the Revised 2016 ACR-EULAR Classification Criteria. *Medicine* **2019**, *98*, e14641. [CrossRef]
49. Quintana, R.; Silvestre, A.M.R.; Goñi, M.; García, V.; Mathern, N.; Jorfen, M.; Miljevic, J.; Dhair, D.; Laithe, M.; Conti, S.; et al. Prevalence of Musculoskeletal Disorders and Rheumatic Diseases in the Indigenous Qom Population of Rosario, Argentina. *Clin. Rheumatol.* **2016**, *35*, 5–14. [CrossRef]
50. Angeles Aguirre, M.; Alarcon, G.; Alegre, J.J.; Alperi, M.; Andreu, J.L. Manual SER de Diagnóstico y Tratamiento de las Enfermedades Reumáticas Autoinmunes Sistémicas. *Soc. Esp. Reumatol. SER* **2014**, 478.
51. Lee, A.J.; Lee, J.; Saw, S.-M.; Gazzard, G.; Koh, D.; Widjaja, D.; Tan, D.T.H. Prevalence and Risk Factors Associated with Dry Eye Symptoms: A Population Based Study in Indonesia. *Br. J. Ophthalmol.* **2002**, *86*, 1347–1351. [CrossRef]
52. Caffery, B.E.; Richter, D.; Simpson, T.; Fonn, D.; Doughty, M.; Gordon, K. CANDEES. The Canadian Dry Eye Epidemiology Study. *Adv. Exp. Med. Biol.* **1998**, *438*, 805–806.
53. Begley, C.G.; Caffery, B.; Nichols, K.; Mitchell, G.L.; Chalmers, R.; DREI Study Group. Results of a Dry Eye Questionnaire from Optometric Practices in North America. *Adv. Exp. Med. Biol.* **2002**, *506*, 1009–1016. [CrossRef] [PubMed]
54. Patel, R.; Shahane, A. The Epidemiology of Sjögren's Syndrome. *Clin. Epidemiol.* **2014**, *6*, 247–255. [CrossRef] [PubMed]
55. The Epidemiology of Dry Eye Disease: Report of the Epidemiology Subcommittee of the International Dry Eye WorkShop (2007). *Ocul. Surf.* **2007**, *5*, 93–107. [CrossRef]
56. Lendrem, D.; Mitchell, S.; McMeekin, P.; Bowman, S.; Price, E.; Pease, C.T.; Emery, P.; Andrews, J.; Lanyon, P.; Hunter, J.; et al. Health-Related Utility Values of Patients with Primary Sjögren's Syndrome and Its Predictors. *Ann. Rheum. Dis.* **2014**, *73*, 1362–1368. [CrossRef] [PubMed]
57. Hackett, K.L.; Newton, J.L.; Frith, J.; Elliott, C.; Lendrem, D.; Foggo, H.; Edgar, S.; Mitchell, S.; Ng, W.-F. Impaired Functional Status in Primary Sjögren's Syndrome. *Arthritis Care Res.* **2012**, *64*, 1760–1764. [CrossRef] [PubMed]
58. Mejía, L.F.; Gil, J.C.; Jaramillo, M. Intense Pulsed Light Therapy: A Promising Complementary Treatment for Dry Eye Disease. *Arch. Soc. Espanola Oftalmol.* **2019**, *94*, 331–336. [CrossRef]
59. Liu, K.C.; Huynh, K.; Grubbs, J.; Davis, R.M. Autoimmunity in the Pathogenesis and Treatment of Keratoconjunctivitis Sicca. *Curr. Allergy Asthma Rep.* **2014**, *14*, 403. [CrossRef]
60. Zhou, L.; Wei, R.; Zhao, P.; Koh, S.K.; Beuerman, R.W.; Ding, C. Proteomic Analysis Revealed the Altered Tear Protein Profile in a Rabbit Model of Sjögren's Syndrome-Associated Dry Eye. *Proteomics* **2013**, *13*, 2469–2481. [CrossRef]
61. Vitali, C.; Bombardieri, S.; Jonsson, R.; Moutsopoulos, H.M.; Alexander, E.L.; Carsons, S.E.; Daniels, T.E.; Fox, P.C.; Fox, R.I.; Kassan, S.S.; et al. Classification Criteria for Sjögren's Syndrome: A Revised Version of the European Criteria Proposed by the American-European Consensus Group. *Ann. Rheum. Dis.* **2002**, *61*, 554–558. [CrossRef]
62. Romão, V.C.; Talarico, R.; Scirè, C.A.; Vieira, A.; Alexander, T.; Baldini, C.; Gottenberg, J.-E.; Gruner, H.; Hachulla, E.; Mouthon, L.; et al. Sjögren's Syndrome: State of the Art on Clinical Practice Guidelines. *RMD Open* **2018**, *4*, e000789. [CrossRef]
63. Hay, E.M.; Thomas, E.; Pal, B.; Hajeer, A.; Chambers, H.; Silman, A.J. Weak Association between Subjective Symptoms or and Objective Testing for Dry Eyes and Dry Mouth: Results from a Population Based Study. *Ann. Rheum. Dis.* **1998**, *57*, 20–24. [CrossRef] [PubMed]
64. Tseng, S.C. Evaluation of the Ocular Surface in Dry-Eye Conditions. *Int. Ophthalmol. Clin.* **1994**, *34*, 57–69. [CrossRef] [PubMed]
65. Begley, C.G.; Caffery, B.; Chalmers, R.L.; Mitchell, G.L.; Dry Eye Investigation (DREI) Study Group. Use of the Dry Eye Questionnaire to Measure Symptoms of Ocular Irritation in Patients with Aqueous Tear Deficient Dry Eye. *Cornea* **2002**, *21*, 664–670. [CrossRef]
66. McMonnies, C.W.; Ho, A. Responses to a Dry Eye Questionnaire from a Normal Population. *J. Am. Optom. Assoc.* **1987**, *58*, 588–591. [PubMed]

67. Baer, A.N.; Walitt, B. Update on Sjögren Syndrome and Other Causes of Sicca in Older Adults. *Rheum. Dis. Clin. N. Am.* **2018**, *44*, 419–436. [CrossRef] [PubMed]
68. Roszkowska, A.M.; Oliverio, G.W.; Aragona, E.; Inferrera, L.; Severo, A.A.; Alessandrello, F.; Spinella, R.; Postorino, E.I.; Aragona, P. Ophthalmologic Manifestations of Primary Sjögren's Syndrome. *Genes* **2021**, *12*, 365. [CrossRef]
69. Vivino, F.B.; Carsons, S.E.; Foulks, G.; Daniels, T.E.; Parke, A.; Brennan, M.T.; Forstot, S.L.; Scofield, R.H.; Hammitt, K.M. New Treatment Guidelines for Sjögren's Disease. *Rheum. Dis. Clin. N. Am.* **2016**, *42*, 531–551. [CrossRef]
70. Bron, A.J.; Evans, V.E.; Smith, J.A. Grading of Corneal and Conjunctival Staining in the Context of Other Dry Eye Tests. *Cornea* **2003**, *22*, 640–650. [CrossRef]
71. Chalmers, R.L.; Begley, C.G.; Caffery, B. Validation of the 5-Item Dry Eye Questionnaire (DEQ-5): Discrimination across Self-Assessed Severity and Aqueous Tear Deficient Dry Eye Diagnoses. *Contact Lens Anterior Eye J. Br. Contact Lens Assoc.* **2010**, *33*, 55–60. [CrossRef]
72. Lin, P.-Y.; Cheng, C.-Y.; Hsu, W.-M.; Tsai, S.-Y.; Lin, M.-W.; Liu, J.-H.; Chou, P. Association between Symptoms and Signs of Dry Eye among an Elderly Chinese Population in Taiwan: The Shihpai Eye Study. *Investig. Ophthalmol. Vis. Sci.* **2005**, *46*, 1593–1598. [CrossRef]
73. Dogru, M.; Tsubota, K. New Insights into the Diagnosis and Treatment of Dry Eye. *Ocul. Surf.* **2004**, *2*, 59–75. [CrossRef]
74. Zeev, M.S.-B.; Miller, D.D.; Latkany, R. Diagnosis of Dry Eye Disease and Emerging Technologies. *Clin. Ophthalmol.* **2014**, *8*, 581–590. [CrossRef] [PubMed]
75. Sullivan, B.D.; Crews, L.A.; Messmer, E.M.; Foulks, G.N.; Nichols, K.K.; Baenninger, P.; Geerling, G.; Figueiredo, F.; Lemp, M.A. Correlations between Commonly Used Objective Signs and Symptoms for the Diagnosis of Dry Eye Disease: Clinical Implications. *Acta Ophthalmol.* **2014**, *92*, 161–166. [CrossRef] [PubMed]
76. Mengher, L.S.; Pandher, K.S.; Bron, A.J. Non-Invasive Tear Film Break-up Time: Sensitivity and Specificity. *Acta Ophthalmol.* **1986**, *64*, 441–444. [CrossRef]
77. Giannaccare, G.; Scorcia, V. False Myths versus Medical Facts: Ten Common Misconceptions Related to Dry Eye Disease. *Biomedicines* **2020**, *8*, 172. [CrossRef] [PubMed]
78. Acs, M.; Caffery, B.; Barnett, M.; Edmonds, C.; Johnson-Tong, L.; Maharaj, R.; Pemberton, B.; Papinski, D.; Harthan, J.; Srinivasan, S. Customary Practices in the Monitoring of Dry Eye Disease in Sjogren's Syndrome. *J. Optom.* **2018**, *11*, 232–241. [CrossRef]
79. Shen Lee, B.; Kabat, A.G.; Bacharach, J.; Karpecki, P.; Luchs, J. Managing Dry Eye Disease and Facilitating Realistic Patient Expectations: A Review and Appraisal of Current Therapies. *Clin. Ophthalmol.* **2020**, *14*, 119–126. [CrossRef]
80. Solans-Laqué, R.; López-Hernandez, A.; Bosch-Gil, J.A.; Palacios, A.; Campillo, M.; Vilardell-Tarres, M. Risk, Predictors, and Clinical Characteristics of Lymphoma Development in Primary Sjögren's Syndrome. *Semin. Arthritis Rheum.* **2011**, *41*, 415–423. [CrossRef]
81. Lazarus, M.N.; Robinson, D.; Mak, V.; Møller, H.; Isenberg, D.A. Incidence of Cancer in a Cohort of Patients with Primary Sjogren's Syndrome. *Rheumatol. Oxf. Engl.* **2006**, *45*, 1012–1015. [CrossRef]
82. Suresh, L.; Malyavantham, K.; Shen, L.; Ambrus, J.L. Investigation of Novel Autoantibodies in Sjogren's Syndrome Utilizing Sera from the Sjogren's International Collaborative Clinical Alliance Cohort. *BMC Ophthalmol.* **2015**, *15*, 38. [CrossRef]
83. Barboza, M.N.C.; Barboza, G.N.C.; de Melo, G.M.; Sato, E.; Dantas, M.C.N.; Dantas, P.E.C.; Felberg, S. [Correlation between signals and symptoms of dry eye in Sjögren's syndrome patients]. *Arq. Bras. Oftalmol.* **2008**, *71*, 547–552. [CrossRef] [PubMed]
84. Figueroa-Ortiz, L.C.; Jiménez Rodríguez, E.; García-Ben, A.; García-Campos, J. [Study of tear function and the conjunctival surface in diabetic patients]. *Arch. Soc. Espanola Oftalmol.* **2011**, *86*, 107–112. [CrossRef] [PubMed]
85. Chaw, P.F.A.; del Carmen Arteaga Loor, G.; Vera, M.C.M.; Fernández, A.C.S. Los factores de riesgo para el síndrome del ojo seco en la población manabita. *Cienc. Educ. Rev. Científica* **2020**, *1*, 56–64. [CrossRef]
86. Rodríguez, A.M.; Rojas, B.A.; Mercado, M. Asociación entre los síntomas y signos en pacientes con ojo seco. *Cienc. Tecnol. Salud Vis. Ocul.* **2008**, *6*, 12.
87. Singh, S.; Srivastav, S.; Modiwala, Z.; Ali, M.H.; Basu, S. Repeatability, reproducibility and agreement between three different diagnostic imaging platforms for tear film evaluation of normal and dry eye disease. *Eye*. [CrossRef]
88. Sánchez-González, M.C.; Capote-Puente, R.; García-Romera, M.-C.; De-Hita-Cantalejo, C.; Bautista-Llamas, M.-J.; Silva-Viguera, C.; Sánchez-González, J.-M. Dry eye disease and tear film assessment through a novel non-invasive ocular surface analyzer: The OSA protocol. *Front. Med.* **2022**, *9*, 938484. [CrossRef] [PubMed]
89. Tian, L.; Qu, J.H.; Zhang, X.Y.; Sun, X.G. Repeatability and reproducibility of noninvasive keratograph 5m measurements in patients with dry eye disease. *J. Ophthalmol.* **2016**, *2016*, 8013621. [CrossRef]
90. Itokawa, T.; Suzuki, T.; Iwashita, H.; Hori, Y. Comparison and evaluation of prelens tear film stability by different noninvasive in vivo methods. *Clin. Ophthalmol.* **2020**, *14*, 4459–4468. [CrossRef]
91. Borroni, D.; Paytuví-Gallart, A.; Sanseverino, W.; Gómez-Huertas, C.; Bonci, P.; Romano, V.; Giannaccare, G.; Rechichi, M.; Meduri, A.; Oliverio, G.W.; et al. Exploring the Healthy Eye Microbiota Niche in a Multicenter Study. *Int. J. Mol. Sci.* **2022**, 23. [CrossRef]
92. Deng, C.; Xiao, Q.; Fei, Y. A Glimpse into the Microbiome of Sjögren's Syndrome. *Front. Immunol.* **2022**, *13*, 918619. [CrossRef]
93. Watane, A.; Raolji, S.; Cavuoto, K.; Galor, A. Microbiome and immune-mediated dry eye: A review. *BMJ Open Ophthalmol.* **2022**, *7*, e000956. [CrossRef]
94. De Luca, F.; Shoenfeld, Y. The microbiome in autoimmune diseases. *Clin. Exp. Immunol.* **2019**, *195*, 74–85. [CrossRef] [PubMed]
95. Cavuoto, K.M.; Banerjee, S.; Galor, A. Relationship between the microbiome and ocular health. *Ocul. Surf.* **2019**, *17*, 384–392. [CrossRef] [PubMed]

life

MDPI

Article

Accommodation Response Variations in University Students under High Demand for Near-Vision Activity

Concepción De-Hita-Cantalejo, María-de-los-Ángeles Benítez-Rodríguez, María Carmen Sánchez-González, María-José Bautista-Llamas and José-María Sánchez-González *

Department of Physics of Condensed Matter, Optics Area, Vision Science Research Group (CIVIUS), University of Seville, 41004 Seville, Spain
* Correspondence: jsanchez80@us.es

Abstract: The objective of this study was to investigate accommodation changes and visual discomfort in a university student population after a period of high demand for near-vision activity. A total of 50 university students aged between 20 and 22 years were recruited. The tests performed involved positive relative accommodation (PRA), negative relative accommodation (NRA), accommodation amplitude (AA), and monocular and binocular accommodative facility (MAF and BAF). Visual discomfort was measured on a scale involving a visual discomfort questionnaire (VDQ). All accommodative variables underwent changes during the exam period; specifically, regarding NRA and PRA, 30.4% and 15.1% of the studied population, respectively, appeared to be below average. Moreover, 42.3% of the population exhibited values below average in the second measure of AA. On the other hand, a small percentage of the population was below average in MAF and BAF measurements: 3% in the monocular right eye test, 6% in the left eye test, and 9.1% in the binocular facility test. Finally, the VDQ score did not reveal a statistically significant difference between the two measurements. Prolonged near-distance work, such as a university exams period, changed all accommodation systems (amplitude of accommodation, relative accommodation, and accommodation facility). These changes influence an accommodation excess that results in blurred vision, headache, and problems with focusing.

Keywords: accommodation response; university population; near-vision activities; accommodation disorders

Citation: De-Hita-Cantalejo, C.; Benítez-Rodríguez, M.-d.-l.-Á.; Sánchez-González, M.C.; Bautista-Llamas, M.-J.; Sánchez-González, J.-M. Accommodation Response Variations in University Students under High Demand for Near-Vision Activity. *Life* **2022**, *12*, 1837. https://doi.org/10.3390/life12111837

Academic Editor: Zdenka Bendova

Received: 17 October 2022
Accepted: 8 November 2022
Published: 9 November 2022

1. Introduction

The human eye has the ability to adapt to different distances via changes in its refractive power, thus maintaining a clear image focused on the retina [1]. Nevertheless, accommodative response is flawed as a result of prolonged effort in near vision [2,3], resulting in accommodative and nonstrabismic binocular dysfunctions that affect the patient's visual performance, especially during near-vision activities [4,5]. Due to excessive effort, the visual system loses efficacy and results in symptoms such as headache, blurred vision, or difficulty in focusing [6,7]. Different criteria have been used to diagnose these visual disturbances, either by observing the presence or absence of symptoms, using a symptomatology questionnaire, or carrying out an assessment test of the accommodative system [4,8]. Accommodative dysfunctions are among the most frequent causes of asthenopia symptomatology [9,10]; therefore, evaluation of the accommodative system is of great importance for the prevention of such issues. Assessment tools that provide sufficient information on the accommodative function include monocular accommodative amplitude (AA) [11,12] and monocular accommodative facility (AF) in both the phase with negative lenses and the phase with positive lenses [13]. Indirectly, by assessing binocular AA, negative and positive relative accommodation (NRA and PRA) and binocular AF are shown in both phases [8].

Negative relative accommodation (NRA) and positive relative accommodation (PRA) measure the maximum capacity to stimulate accommodation while maintaining unique binocular vision and indirectly provide information on fusional vergence [14,15]; therefore, determining these values aids in the diagnosis of accommodative and vergence-related problems. A low PRA value indicates accommodative insufficiency, and a low NRA value reveals accommodative excess [13,16]. The amplitude of accommodation (AA) is the maximum focusing power of the eye, measured in diopters [17]. There are different methods to perform this measurement; however, we needed to determine the most reproducible and reliable method that could detect statistically significant changes in the measured values [18]. Antona et al. conducted a study comparing different measuring methods of AA, concluding that the most repeatable method was that of negative lenses [19].

Monocular accommodative facility (MAF) measures the ability of the eye to vary its accommodative state by changing focus for a certain time, accurately and repeatedly [20]. This test is performed as a measure of visual fatigue that can be caused by accommodative dysfunctions [13]. Values outside the norm are indicative of accommodative alterations, and some authors have linked these alterations with decreased academic performance [4,21]. Therefore, most studies have focused more on children rather than adults. In addition, there are no unified criteria on clinical signs to diagnose these alterations [22]. Although some studies have obtained data on visual symptomatology in university students [23], there are a lack of studies analyzing the negative effects of excessive use of near vision. Due to the high visual demand for several hours [24], it has been observed that near-point measurements after near work are predictive of symptoms in students, which shows that a responsible mechanism produces this symptomatology after a continuous effort in near vision [25].

Visual discomfort was measured on a scale using a visual discomfort questionnaire (VDQ). This scale was considered valid and reliable for assessing the symptomatology [26].

Consequently, the purpose of this study was to investigate accommodative changes in a population of university students after an exam period, when there is a high demand for near-vision activities.

2. Materials and Methods

2.1. Design

A total of 50 patients were recruited for the study. All participants were students aged between 20 and 22 years who were given a two-stage accommodative test, the first part just after a holiday period, in September 2019, and the second part after an exam period in the first quarter, in January 2020. All tests were performed on the maximum positive refraction to result in the best visual acuity (VA) of each patient.

2.2. Ethical Aspects

All patients included in this work were adequately informed verbally and in writing of the tests to be performed on them. All patients signed an informed consent form prior to the start of the study. This study was conducted in accordance with the Declaration of Helsinki, and the Institutional Review Board of the University Hospital Virgen Macarena of the University of Seville approved the research.

2.3. Subjects

The 50 patients voluntarily attended the University of Seville. The inclusion criteria were as follows: (1) university students between the ages of 20 and 22 years, and (2) VA with or without correction of 0.0 logMAR or higher. The exclusion criteria were: (1) subjects with binocular dysfunctions, strabismus, nystagmus, or amblyopia; (2) persons with intellectual disabilities; (3) subjects who did not sign the informed consent form; (4) subjects who did not complete the survey; and (5) subjects who had undergone some type of ocular surgery.

Of the initial enrolled population, 13 students were excluded for having binocular problems. Of the 37 remaining candidates who enrolled in the study, 4 were excluded for not completing the tests. Therefore, a total of 33 students' data were analyzed.

2.4. Materials and Measurements

The measurements were carried out in the laboratory of the Faculty of Pharmacy of the University of Seville. The variables with which the accommodative function was measured were: accommodative amplitude (diopters, D); negative and positive relative accommodation (D), evaluated with the ESSILOR MPH100E S/N 000104 phoropter (Essilor, Paris, France); and accommodative facility (cycles per minute), quantified with ±2 flipper lenses (Optometric Promotion, Burgos, Spain).

Before starting each test, the procedure to be performed was explained to the patient. Three measurements were made. The mean value of the three values obtained was calculated. As far as testing and regulatory values are concerned, we relied on the manual by Scheiman and Wick [8].

For the measurement of PRA, the ability to stimulate accommodation was evaluated by adding negative lenses; in the case of NRA, the ability to relax accommodation was evaluated using positive lenses. For the measurements, we used a phoropter with a correction in far vision for each patient. The optotype was located about 40 cm from the eye, and IPD was near distance. The subject was instructed to look at a line lower than that of their VA on the view card. Negative lenses were introduced for the measurement of PRA and positive lenses for the measurement of NRA, binocularly in steps of 0.25 D until the first point of blurriness occurred. The normal values set for NRA were +2 × 0.50 D and for PRA, −2.37 × 0.50 D [8].

The AA test was performed using the negative lens method. The optotype was placed in the phoropter at 40 cm with a near IPD. Negative lenses were inserted binocularly in steps of −0.25 D until the patient noticed a sustained blur, i.e., a continuous blurring. The value of AA corresponded to the power of the last lens with which the subject was able to clarify the optotype, adding the necessary accommodations to see at a 40 cm distance, i.e., 2.50 D. For the normal values, Hofstetter created three equations of AA that present its variation with the progression of age [27]. These three involve the minimum, the mean, and the maximum. In this study, the equation with the mean was used, which corresponds to the formula AA media $-18.5 - 0.3 \times$ age (years) [8]. All optometry instruments were from Optometric Promotion (Burgos, Spain)

With the MAF test, we evaluated the ability of the accommodative system to make rapid and abrupt accommodative changes, thus checking the fatigue resistance in each time and distance. During the measurements, we used a manual near card at a 40 cm distance from the patient. For both monocular (MAF) and binocular (BAF) tests, the patient was asked to look at a lower line than that of their maximum VA in the near-up optotype, which must always remain clear. The flipper was introduced with the positive lens, and we expected the participants to declare sharp vision before we turned the flipper to the negative lens, as well as the same observation from the subjects. The same process was followed for the other eye and, finally, binocularly. At the same time, we recorded the cycles per minute (cpm). The normal values set were 11 ± 5 cpm for MAF and 8 ± 5 cpm for BAF [8].

In addition, the variables that were measured to rule out binocular dysfunctions were: the magnitude of the horizontal heterophoria (Prism Diopters, Δ) [28] and the amplitude of both the positive (convergence) and negative fusional vergences (divergence) [29]. According to the value of these variables and Sheard's criterion [30], subjects with binocular dysfunctions were excluded. Sheard's criterion was defined as follows: in order to have a comfortable binocular vision, the value of the fusional vergence must be twice the value of the phoria.

To evaluate visual discomfort, we used the scale of visual discomfort questionnaire (VDQ), with questions regarding study time, study breaks, headache, eye pain, focus issues, blurry vision, and double vision. The VDQ [31] consisted of 23 items with a four-point scale: 0 = event never occurs; 1 = occasionally, a couple of times a year; 2 = often, every few weeks; and 3 = almost always. The items proposed by Borsting et al. [31] were well within the reading level of college students. The items of the VDQ are presented in Table 1.

Table 1. Item list of the Visual Discomfort Questionnaire by Borsting et al. [31].

Question	Item
When reading, do the words or letters in the words ever appear to spread apart?	movement/fading
Do you ever have difficulty reading the words on a page because they begin to flicker or shimmer?	movement/fading
Does the white background behind the text ever appear to move, flicker, or shimmer, making the letters hard to read?	movement/fading
Do you ever have difficulty seeing more than one or two words on a line in focus?	blur/diplopia
When reading, do the words on the page ever begin to move or float?	movement/fading
When you are reading a page that consists of black print on white background, does the background ever appear to overtake the letters, making them hard to read?	movement/fading
When reading, do the words on a page of clear text ever appear to fade into the background, then reappear?	movement/fading
Do the letters on a page ever appear as a double image when you are reading?	blur/diplopia
When reading black print on a white background, do you ever have to move the page around or continually blink to avoid glare that seems to come from the background?	movement/fading
Do you ever get a headache from reading a newspaper or magazine with clear print?	headache
Do you have to move your eyes around the page or continually blink or rub your eyes to keep the text easy to see when you are reading?	blur/diplopia
When reading, do you ever have to squint to keep the words on a page of clear text from going blurry or out of focus?	blur/diplopia
Do the letters on a page of clear text ever go blurry when you are reading?	blur/diplopia
When reading, do you ever have difficulty keeping the words on the page of clear text in focus?	blur/diplopia
Do your eyes ever feel watery, red, sore, strained, tired, dry, or gritty after you have been reading a newspaper or magazine with clear print?	sore
Do your eyes ever feel watery, red, sore, strained, tired, dry, or gritty, or do you rub them a lot, when viewing a striped pattern?	sore
Do you have to use a pencil or your finger to keep from losing your place when reading a page of text in a novel or magazine?	rereading
As a result of any of the above difficulties, do you find reading a slow task?	rereading
How often do you get a headache when working under fluorescent lights?	headache
Do your eyes ever feel watery, red, sore, strained, tired, dry, or gritty, when working under fluorescent lights?	sore
When reading under fluorescent lights or in bright sunlight, does the glare from the bright white glossy pages cause you to continually move the page around so that you can see the words clearly?	glare
When reading, do you ever unintentionally re-read the same words in a line of text?	rereading
When reading, do you ever unintentionally re-read the same line?	rereading

2.5. Data Analysis

Statistical analysis was carried out with SPSS statistics 25.0 (IBM Corporation, Armonk, NY, USA). All visual acuity data were converted into Snellen formats. The Student's *t*-test was applied for parametric-dependent variables. Effect size calculation was assessed within D of the Cohen test [32]. All statistical tests were performed with a 95% confidence interval ($p < 0.05$).

3. Results

3.1. Relative Accommodation

The pre-exams score for NRA was +2.59 ± 0.58 (+1.25 to +4.00) diopters, whereas the post-exams NRA score was +1.69 ± 0.51 (+0.75 to +3.00) diopters. Therefore, NRA was characterized by statistically significant differences between the two measurements (mean difference, 0.89 ± 0.59 diopters; 95% confidence interval, 0.68 to 1.10; t = 8.64;

p = 0.007) (Figure 1A). Cohen´s D size was 1.64, which is considered a large effect size. The normative values for NRA, as established by Scheiman and Wick [8], were +2.00 ± 0.50 D. Our results indicated that before the exam period, 3% of the studied population were under the norm, 42.5% were within the norm, and 54.5% ere above the norm. After the exam period, 30.4% were under the norm, 63.6% were within the norm, and only 6% were above the normative values. The PRA score obtained before the exam period was −1.66 ± 1.00 (−0.50 to −5.00) diopters and reached −2.36 ± 1.01 (−0.25 to −4.50) diopters after the exam period. Therefore, PRA exhibited statistically significant differences between the two measurements (mean difference, 0.69 ± 1.10 diopters; 95% confidence interval, 0.30 to 1.08; t = 3.62, p = 0.009) (Figure 1B). Cohen´s D size was 0.69, which is considered a medium effect size. The normative values for PRA, as established by Scheiman and Wick [8], were −2.37 ± 1.00 D. Our results revealed that before the exam period, 48.5% of the studied population were under the norm, 45.5% were within the norm, and 6% were above the norm. Post-exams, 15.1% were under the norm, 72.8% were within the norm, and only 12.1% were above the normative values.

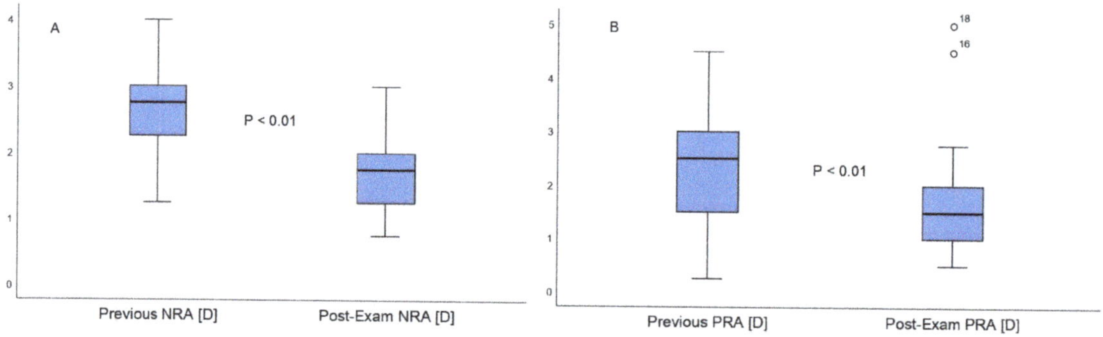

Figure 1. Negative Relative Accommodation (**A**) and Positive Relative Accommodation (**B**) comparative boxplots between the first and second measurement, respectively.

3.2. *Amplitude of Accommodation*

The pre-exams score for AA was 6.70 ± 1.34 (2.00 to 9.25) diopters and reached 7.24 ± 1.53 (3.75 to 9.75) diopters after the exam period. Therefore, AA exhibited statistically significant differences between the two measurements (mean difference, 0.53 ± 1.38 diopters; 95% confidence interval, 0.05 to 1.02; t = 2.23, p = 0.03) (Figure 2). Cohen´s D size was 0.37, which is considered a medium effect size. The normative values for AA, as established by Scheiman and Wick [8], were $16 - {}^{1}/_{3}$ age ± 2.00 D (mean age was used, 8.99 ± 2.00 D). Our results demonstrated that before the exam period, 60.5% of the studied population were under the norm, and 39.5% were within the norm. Post-exams, 42.3% were under the norm, and 57.7% were within the norm. We did not observe AA scores above the normative values either pre- or post-exams.

3.3. *Accommodative Facility*

The right eye MAF score obtained before the exam period was 13.24 ± 5.36 (3.00 to 25.00) cpm and reached 11.30 ± 4.44 (4.00 to 26.00) cpm after the exam period. Therefore, the right eye MAF reported statistically significant differences between the two measurements (mean difference, 1.93 ± 4.83 cpm; 95% confidence interval, 0.22 to 3.65; t = 2.23, p = 0.02) (Figure 3A), with a Cohen´s D size of 0.39, which is considered a medium effect size. The right eye MAF normative values, as established by Scheiman and Wick [8], were 11 ± 5 cpm. Our results reported that before the exam period, 6.1% of the studied population were under the norm, 66.8% were within the norm, and 27.2% were above the norm. After the exam period, 3% were under the norm, 88% were within the norm, and 9% were above the normative values. The left eye MAF score obtained before the exam period was 14.27 ± 4.95 (2.00 to 25.00) cpm and decreased to 11.42 ± 5.20 (1.00 to 28.00) cpm post-exams. Therefore, the left eye MAF differences between the two measurements were statistically significant (mean difference, 2.84 ± 4.61 cpm; 95% confidence interval, 1.21 to 4.48; t = 3.54, p = 0.009) (Figure 3B), with a Cohen´s D size of 0.56, which is considered a medium effect size. The left eye MAF normative values, as established by Scheiman and Wick [8], were 11 ± 5 cpm. Our results revealed that before the exam period, 6% of the studied population were under the norm, 63.7% were within the norm, and 30.3% were above the norm. Post-exams, 6% were under the norm, 82% were within the norm, and 12% were above the normative values. Additionally, the pre-exams BAF score was 13.51 ± 4.36 (4.00 to 20.00) cpm and decreased to 10.96 ± 4.53 (3.00 to 25.00) cpm after the exam period. Therefore, BAF was also characterized by statistically significant differences between the two measurements (mean difference, 2.54 ± 4.72 cpm; 95% confidence interval, 0.82 to 4.22; t = 3.09, p = 0.004) (Figure 3C), with a Cohen´s D size of 0.57, which is regarded as a medium effect size. The BAF normative values, as established by Scheiman and Wick [8], were 10 ± 5 cpm. Our results demonstrated that before the exam period, 3% of the studied population were under the norm, 57.6% were within the norm, and 39.4% were above the norm. Post-exams, 9.1% were under the norm, 78.9% were within the norm, and 12% were above the normative values.

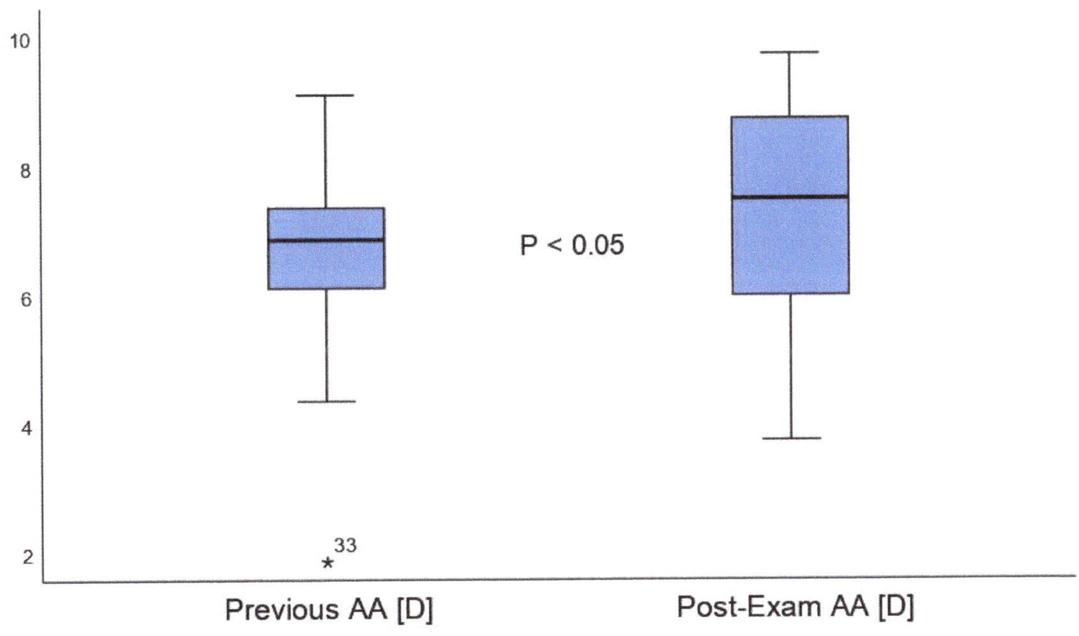

Figure 2. Accommodation Amplitude comparative boxplots between the first and second measurement.

3.4. Visual Discomfort

Finally, the visual discomfort questionnaire (VDQ) score obtained before the exam period was 17.84 ± 1.08 (4.00 to 31.00) and 17.93 ± 0.96 (6.00 to 28.00) post-exams. Therefore, the VDQ score did not present statistically significant differences between the two measurements (mean difference, −0.09 ± 3.80 points; 95% confidence interval, −1.43 to 1.25; t = −0.13, p = 0.892), with a Cohen´s D size of 0.08, which is regarded as a small effect size. No further statistical analysis related to VDQ score was conducted.

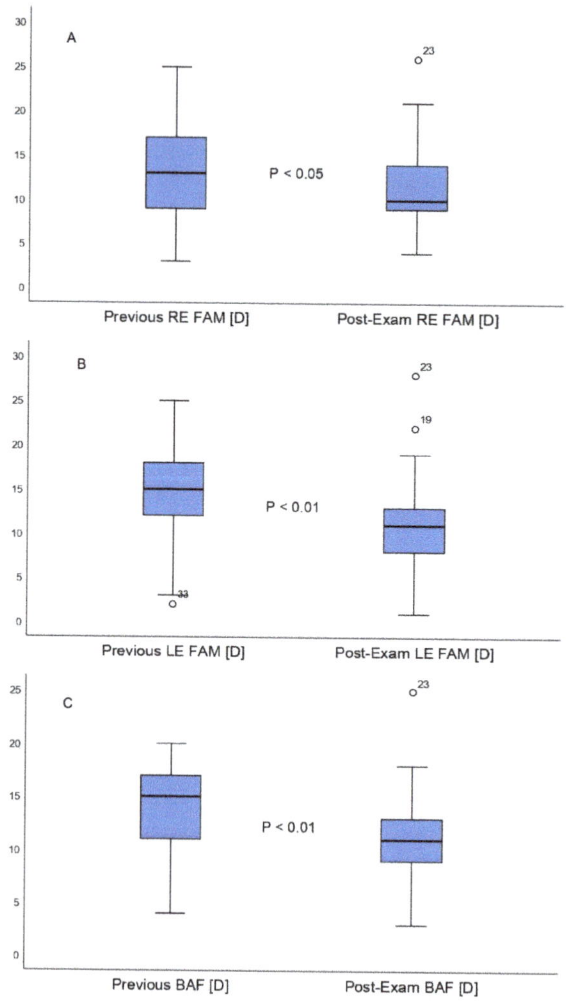

Figure 3. Right Eye Monocular Accommodative Facility (**A**), Left Eye Monocular Accommodative Facility (**B**), and Binocular Accommodative Facility (**C**) comparative boxplots between the first and second measurement, respectively.

4. Discussion

The results of our study on young university students indicated that all the accommodative variables measured underwent changes after the exam period. Measures of relative accommodation, both positive and negative, accommodative facility, and amplitude of accommodation were carried out. These tests have been used by several authors to identify the type of disorder [33,34]. This system developed by Scheiman and Wick is commonly used as a reference for the classification, diagnosis, and treatment of accommodative disorders [8].

The mean NRA was (+2.59 ± 0.58) D pre-exams and (+1.69 ± 0.51) D post-exams. In addition, the measure PRA was (−1.66 ± 1.00) D pre-exams and (−2.36 ± 1.01) D post-exams in our study, respectively. Negative relative accommodation (NRA) and positive relative accommodation (PRA) are used as diagnostic tests of accommodative disorders. Many authors establish a relationship between low NRA and PRA values and accommodative excess and insufficiency, respectively [24]. García et al. [35] determined that high PRA values do relate to disorders associated with accommodative excess, and they considered this alteration as a diagnostic measure of this anomaly. A low value of PRA indicates that the patient does not admit the introduction of negative lenses. The subject is not able to increase his/her accommodation to a situation that may be related to an accommodative insufficiency. A low value NRA indicates that the subject does not admit the binocular interposition of positive lenses, because he is not able to relax the accommodation; it probably indicates accommodative excess.

Other studies [11,36] on university students in which measures of accommodative parameters were taken have observed significant changes in PRA, which were associated with accommodative dysfunctions. The same observation was made [37] in young workers who perform daily near-vision work (high PRA values after increased near work). Thus, the variation in PRA values coincides with the data in our study collected in the second measurement. Although our values are lower, they could represent the beginning of an accommodative excess in the study population.

Concerning AA, the results of the present work seem to be in the same line as those obtained by different authors who concluded that AA increases after many hours of work in near vision [22,38]. The mean AA was (6.70 $\pm$ 1.34) pre-exams and (7.24 $\pm$ 1.53) D post-exams. Considering different studies, we decided to use the low accommodative amplitude and high accommodation amplitude as the clinical sign for diagnosing accommodative insufficiency and excess [11]. Literature-reported prevalence of accommodative disorders varies greatly due to the lack of standardization in the type of subjects enrolled and clinical diagnostic tests employed. However, our results are in agreement with other studies that state that increased near-visual activity increases AA and produces an accommodative excess [39].

Further, to determine the state of the accommodative function, it is necessary to assess the monocular accommodative facility (AF) both in the phase with positive lenses and in the phase with negative lenses, as well as binocular AF in both phases. The fundamental clinical sign for accommodative excess was failing monocular accommodative facility with positive lenses.

MAF values were also modified, although to a lesser extent. MAF values were very similar before and after the exam period, with 27.2% of the right eye (OD) and 30.3% of the left eye (OI) above the norm before the exam period, 3% of the OD and 6% of the OI below the norm, and 88% of the OD and 82% of the OI within the average after the exam period. BAF results were 57.6% within the norm and 39.4% above the norm before the exam period, and 78.9% within the norm, 12% above the norm, and 9.1% below the norm after the exams. Porcar et al. [39] analyzed the accommodative and binocular dysfunctions caused by the use of computers. They included 89 patients who underwent vergence and accommodative tests; all had been using digital screens an average of 5 $\pm$ 9 h/day, and none was diagnosed with visual system disturbance. Regarding MAF, 24% of the patients failed focusing positive lenses, 8% failed focusing negative lenses, and 3% with both. Based on all the data obtained, they concluded that the patients analyzed were more prone to accommodative excess, especially considering that, currently, students use computers for longer durations. This observation confirms the result of our study.

Finally, the VDQ obtained 17.84 $\pm$ 1.08 (4.00 to 31.00) points before the exam period and 17.93 $\pm$ 0.96 (6.00 to 28.00) post-exams. Therefore, the VDQ did not report statistically significant differences between the two measurements (t = -0.13, $p = 0.892$). Borsting et al. [23] conducted a study in which they assessed symptoms of visual discomfort in 23 university students for one year using the same scale, reporting no statistically significant differences between the first and second measure of their survey. They concluded that the symptomatology was stable in most patients, supporting the data of our study. On the one hand, the significantly subjective measurements changes were connected with blurred vision, headache, and accommodation problems. All accommodation variables became worse after the period exam. However, on the other hand, the objective questionnaire was a self-patient sensation report. The patients do not perceive these accommodation alterations when filling in the questionnaire; therefore, it supposes an element that increases severity. This indicates that they do not notice symptoms, and it is necessary to carry out periodic visual check-ups of the students during exam periods.

This study presents some limitations that should be considered. One of these is the small sample size, as several patients were left out due to the exclusion criteria. Therefore, studies with a larger sample size would be needed. On the other hand, at the second stage of the study, some of the recruited patients did not show up for the second appointment, thus further decreasing the total number of measures. Another limitation is that the studied population included students of a certain age range; consequently, this study should be performed in both adults over 25 years of age and in children.

5. Conclusions

Prolonged near-distance work, such as a university exams period, changes all accommodation systems (amplitude of accommodation, relative accommodation, and accommodation facility). These changes could influence an accommodation excess that results in blurred vision, headache, and problems with focusing.

Author Contributions: Conceptualization, C.D.-H.-C., J.-M.S.-G., M.-J.B.-L., M.-d.-l.-Á.B.-R. and M.C.S.-G.; methodology, C.D.-H.-C., J.-M.S.-G., M.-J.B.-L., M.-d.-l.-Á.B.-R. and M.C.S.-G.; software, C.D.-H.-C., J.-M.S.-G., M.-J.B.-L., M.-d.-l.-Á.B.-R. and M.C.S.-G.; validation, C.D.-H.-C., J.-M.S.-G., M.-J.B.-L., M.-d.-l.-Á.B.-R. and M.C.S.-G.; formal analysis, C.D.-H.-C., J.-M.S.-G., M.-J.B.-L., M.-d.-l.-Á.B.-R. and M.C.S.-G.; investigation, C.D.-H.-C., J.-M.S.-G., M.-J.B.-L., M.-d.-l.-Á.B.-R. and M.C.S.-G.; resources, C.D.-H.-C., J.-M.S.-G., M.-J.B.-L., M.-d.-l.-Á.B.-R. and M.C.S.-G.; data curation, C.D.-H.-C., J.-M.S.-G., M.-J.B.-L., M.-d.-l.-Á.B.-R. and M.C.S.-G.; writing—original draft preparation, C.D.-H.-C., J.-M.S.-G., M.-J.B.-L., M.-d.-l.-Á.B.-R. and M.C.S.-G.; writing—review and editing, C.D.-H.-C., J.-M.S.-G., M.-J.B.-L., M.-d.-l.-Á.B.-R. and M.C.S.-G.; visualization, C.D.-H.-C., J.-M.S.-G., M.-J.B.-L., M.-d.-l.-Á.B.-R. and M.C.S.-G.; supervision, C.D.-H.-C., J.-M.S.-G., M.-J.B.-L., M.-d.-l.-Á.B.-R. and M.C.S.-G.; project administration, C.D.-H.-C., J.-M.S.-G., M.-J.B.-L., M.-d.-l.-Á.B.-R. and M.C.S.-G. All authors have read and agreed to the published version of the manuscript.

Funding: This research received no external funding. The English edition has been subsidized with the help of emerging groups from the seventh own research plan of the University of Seville.

Institutional Review Board Statement: The study was conducted in accordance with the Declaration of Helsinki and approved by the Ethics Committee of the University Hospital Virgen Macarena of the University of Seville.

Informed Consent Statement: Informed consent was obtained from all subjects involved in the study.

Data Availability Statement: The data presented in this study are available on request from the corresponding author. The data are not publicly available due to their containing information that could compromise the privacy of research participants.

Acknowledgments: The authors appreciate the support offered by the members of the Department of Physics of Condensed Matter, Faculty of Physics, University of Seville, with special thanks to Javier Romero-Landa and Clara Conde-Amiano. In addition, the authors also appreciate the technical support offered by the members and facilities of the Faculty of Pharmacy, University of Seville, with special thanks to María Álvarez-de-Sotomayor.

Conflicts of Interest: The authors declare no conflict of interest.

References

1. Dubbelman, M.; Sicam, V.A.D.P.; Van der Heijde, G.L. The shape of the anterior and posterior surface of the aging human cornea. *Vis. Res.* **2006**, *46*, 993–1001. [CrossRef] [PubMed]
2. Jaschinski-Kruza, W. Eyestrain in VDU users: Viewing distance and the resting position of ocular muscles. *Hum. Factors* **1991**, *33*, 69–83. [CrossRef] [PubMed]
3. Chen, A.M.; Borsting, E.J. Near work symptoms and measures of accommodation in children. *Clin. Exp. Optom.* **2022**, 1–6. [CrossRef]

4. Sterner, B.; Gellerstedt, M.; Sjostrom, A. Accommodation and the relationship to subjective symptoms with near work for young school children. *Ophthalmic Physiol. Opt.* **2006**, *26*, 148–155. [CrossRef]
5. Ale Magar, J.B.; Shah, S.; Sleep, M.; Dai, S. Assessment of distance-near control disparity in basic and divergence excess paediatric intermittent exotropia. *Clin. Exp. Optom.* **2022**, 1–4. [CrossRef] [PubMed]
6. Chase, C.; Tosha, C.; Borsting, E.; Ridder, W.H. Visual discomfort and objective measures of static accommodation. *Optom. Vis. Sci.* **2009**, *86*, 883–889. [CrossRef]
7. Mármol-Errasti, E.; Cárdenas-Rebollo, J.M.; Rodán, A.; Pagán-Fernández, E.; Jara-García, L.C.; Palomo-Álvarez, C. Measures of accommodative function in secondary school year 9 and year 13: A 4-year longitudinal study. *Graefes Arch. Clin. Exp. Ophthalmol.* **2022**, 1–8. [CrossRef]
8. Scheiman, M.; Wick, B. (Eds.) Diagnosis and general treatment approach: Diagnostic testing. In *Clinical Management of Binocular Vision: Heterophoric, Accommodative, and Eye Movement Disorders*; Lippincott Williams & Wilkins: Philadelphia, PA, USA, 2008; pp. 3–49, ISBN 0781777844.
9. Hokoda, S.C. General binocular dysfunctions in an urban optometry clinic. *J. Am. Optom. Assoc.* **1985**, *56*, 560–562.
10. Huang, Y.; Li, X.; Wang, C.; Zhou, F.; Yang, A.; Chen, H.; Bao, J. Visual acuity, near phoria and accommodation in myopic children using spectacle lenses with aspherical lenslets: Results from a randomized clinical trial. *Eye Vis.* **2022**, *9*, 33. [CrossRef]
11. García-Muñoz, Á.; Carbonell-Bonete, S.; Cantó-Cerdán, M.; Cacho-Martínez, P. Accommodative and binocular dysfunctions: Prevalence in a randomised sample of university students. *Clin. Exp. Optom.* **2016**, *99*, 313–321. [CrossRef]
12. Meng, C.; Zhang, Y.; Wang, S. Changes in accommodation and convergence function after refractive surgery in myopic patients. *Eur. J. Ophthalmol.* **2022**, 11206721221128992. [CrossRef]
13. Scheiman, M.; Gallaway, M.; Coulter, R.; Reinstein, F.; Ciner, E.; Herzberg, C.; Parisi, M. Prevalence of vision and ocular disease conditions in a clinical pediatric population. *J. Am. Optom. Assoc.* **1996**, *67*, 193–202.
14. Morgan, M.W. Accommodation and its relationship to convergence. *Optom. Vis. Sci.* **1944**, *21*, 183–195. [CrossRef]
15. Grosvenor, T.; Grosvenor, T. Primary care optometry. *Optom. Vis. Sci.* **2007**, *60*, 31–34. [CrossRef]
16. García, A.; Cacho, P.; Lara, F.; Megías, R. The relation between accommodative facility and general binocular dysfunction. *Ophthalmic Physiol. Opt.* **2000**, *20*, 98–104. [CrossRef]
17. Burns, D.H.; Allen, P.M.; Edgar, D.F.; Evans, B.J.W. Sources of error in clinical measurement of the amplitude of accommodation. *J. Optom.* **2019**, *13*, 3–14. [CrossRef]
18. Rouse, M.W.; Borsting, E.; Deland, P.N.; Hyman, L.; Hussein, M.; Solan, H.; Cotter, S.; Grisham, D.; Press, L.; Scheiman, M. Reliability of binocular vision measurements used in the classification of convergence insufficiency. *Optom. Vis. Sci.* **2002**, *79*, 254–264. [CrossRef]
19. Antona, B.; Barra, F.; Barrio, A.; Gonzalez, E.; Sanchez, I. Repeatability intraexaminer and agreement in amplitude of accommodation measurements. *Graefes Arch. Clin. Exp. Ophthalmol.* **2009**, *247*, 121–127. [CrossRef]
20. Zellers, J.A.; Alpert, T.L.; Rouse, M.W. A review of the literature and a normative study of accommodative facility. *J. Am. Optom. Assoc.* **1984**, *55*, 31–37.
21. Simons, H.D.; Grisham, J.D. Binocular anomalies and reading problems. *J. Am. Optom. Assoc.* **1987**, *58*, 578–587.
22. Cacho-Martínez, P.; García-Muñoz, Á.; Ruiz-Cantero, M.T. Is there any evidence for the validity of diagnostic criteria used for accommodative and nonstrabismic binocular dysfunctions? *J. Optom.* **2014**, *7*, 2–21. [CrossRef] [PubMed]
23. Borsting, E.; Chase, C.; Tosha, C.; Ridder, W.H. Longitudinal study of visual discomfort symptoms in college students. *Optom. Vis. Sci.* **2008**, *85*, 992–998. [CrossRef] [PubMed]
24. Porcar, E.; Martinez-Palomera, A. Prevalence of general binocular dysfunctions in a population of university students. *Optom. Vis. Sci.* **1997**, *74*, 111–113. [CrossRef]
25. Wilmer, J.B.; Buchanan, G.M. Nearpoint Phorias After Nearwork Predict ADHD Symptoms in College Students. *Optom. Vis. Sci.* **2009**, *86*, 971–978. [CrossRef] [PubMed]
26. Conlon, E.G.; Lovegrove, W.J.; Chekaluk, E.; Pattison, P.E. Measuring visual discomfort. *Vis. Cogn.* **1999**, *6*, 637–663. [CrossRef]
27. Hofstetter, H.W. The relationship of proximal convergence to fusional and accommodative convergence. *Optom. Vis. Sci.* **1951**, *28*, 300–308. [CrossRef]
28. Hrynchak, P.K.; Herriot, C.; Irving, E.L. Comparison of alternate cover test reliability at near in non-strabismus between experienced and novice examiners. *Ophthalmic Physiol. Opt.* **2010**, *30*, 304–309. [CrossRef]
29. Antona, B.; Barrio, A.; Barra, F.; Gonzalez, E.; Sanchez, I. Repeatability and agreement in the measurement of horizontal fusional vergences. *Ophthalmic Physiol. Opt.* **2008**, *28*, 475–491. [CrossRef]
30. Sheard, C. Zones of ocular comfort. *Am. J. Optom.* **1930**, *7*, 9–25. [CrossRef]
31. Borsting, E.; Chase, C.H.; Ridder, W.H. Measuring visual discomfort in college students. *Optom. Vis. Sci.* **2007**, *84*, 745–751. [CrossRef]
32. Cohen, J. The effect size index: D. In *Statistical Power Analysis for the Behavioral Sciences*; Routledge Academic: Abingdon-on-Thames, UK, 1988.
33. Sánchez-González, M.C.; Gutiérrez-Sánchez, E.; Sánchez-González, J.M.; Rebollo-Salas, M.; Ruiz-Molinero, C.; Jiménez-Rejano, J.J.; Pérez-Cabezas, V. Visual system disorders and musculoskeletal neck complaints: A systematic review and meta-analysis. *Ann. N. Y. Acad. Sci.* **2019**, *1457*, 26–40. [CrossRef]

34. Cacho-Martínez, P.; Cantó-Cerdán, M.; Carbonell-Bonete, S.; García-Muñoz, Á. Characterization of Visual Symptomatology Associated with Refractive, Accommodative, and Binocular Anomalies. *J. Ophthalmol.* **2015**, *2015*, 895803. [CrossRef]
35. García, A.; Cacho, P.; Lara, F. Evaluating relative accommodations in general binocular dysfunctions. *Optom. Vis. Sci.* **2002**, *79*, 779–787. [CrossRef]
36. Jorge, J.; de Almeida, J.B.; Parafita, M.A. Binocular vision changes in university students: A 3-year longitudinal study. *Optom. Vis. Sci.* **2008**, *85*, E999–E1006. [CrossRef]
37. Watten, R.G.; Lie, I.; Birketvedt, O. The influence of long-term visual near-work on accommodation and vergence: A field study. *J. Hum. Ergol.* **1994**, *23*, 27–39.
38. Cacho-Martínez, P.; García-Muñoz, Á.; Ruiz-Cantero, M.T. Do we really know the prevalence of accomodative and nonstrabismic binocular dysfunctions? *J. Optom.* **2010**, *3*, 185–197. [CrossRef]
39. Porcar, E.; Montalt, J.C.; Pons, Á.M.; España-Gregori, E. Symptomatic accommodative and binocular dysfunctions from the use of flat-panel displays. *Int. J. Ophthalmol.* **2018**, *11*, 501–505. [CrossRef]

life

MDPI

Article

Pre-Lens Tear Meniscus Height, Lipid Layer Pattern and Non-Invasive Break-Up Time Short-Term Changes with a Water Gradient Silicone Hydrogel Contact Lens

Raúl Capote-Puente [1], María-José Bautista-Llamas [1], Caterina Manzoni [2] and José-María Sánchez-González [1,*]

[1] Vision Research Group (CIVIUS), Department of Physics of Condensed Matter, Optica Area, University of Seville, 41012 Seville, Spain
[2] Department of Materials Science, Optics and Optometry Area, University of Milano-Bicocca, 20122 Milan, Italy
* Correspondence: jsanchez80@us.es

Simple Summary: In this research, we analyze the tear film layer portion that remains on top of contact lenses. Through a non-invasive measurement technique, the volume of tear, the eye redness, the tear film stability and lipid quality and quantity were assessed in a silicone contact lens that has a variation of the percentage of hydration inside. A total of sixty-two contact lens fittings were evaluated 30 min after placement. Among the results obtained, it was observed that the redness of the eye did not display significant changes. The amount of lipids was slightly reduced. There were no changes in tear volume. The first moment in which the tear broke due to desiccation was lower with the use of the contact lens, and the average time of this tear breakup time increased considerably with the use of this material. Finally, the time of the first eyelid was much higher with the use of the contact lens. In summary, water gradient technology increased the contact lens' previous tear film stability and lengthened blinking time. Lehfilcon A suggests an improvement of tear film instability.

Abstract: To evaluate pre-lens tear film volume, stability and lipid interferometry patterns with a silicone hydrogel water content contact lens, a novel, noninvasive, ocular-surface-analyzer technology was used. A prospective, longitudinal, single-center, self-control study was performed in daily or monthly replacement silicone hydrogel contact lens wearers. A tear film analysis was achieved with the Integrated Clinical Platform (ICP) Ocular Surface Analyzer (OSA) from SBM System. The subjects were reassessed, with the contact lens, after 30 min of wearing to quantify the volume, stability and lipid pattern of the short-term pre-lens tear film. Lipid layer thickness decreased from 2.05 ± 1.53 to 1.90 ± 1.73 Guillon patterns ($p = 0.23$). First pre-lens NIBUT decreased from 5.03 ± 1.04 to 4.63 ± 0.89 s ($p = 0.01$). Mean pre-lens NIBUT significantly increased from 15.19 ± 9.54 to 21.27 ± 11.97 s ($p < 0.01$). Lid opening time significantly increased from 26.36 ± 19.72 to 38.58 ± 21.78 s ($p < 0.01$). The silicone hydrogel contact lens with water gradient technology significantly increased the mean pre-lens NIBUT and lid opening time. Lehfilcon A suggested an improvement in contact lens wearers with tear film instability or decreased subjective symptoms of dry eye disease.

Keywords: pre-lens tear film; lipid pattern; non-invasive break-up time; contact lens

check for updates

Citation: Capote-Puente, R.; Bautista-Llamas, M.-J.; Manzoni, C.; Sánchez-González, J.-M. Pre-Lens Tear Meniscus Height, Lipid Layer Pattern and Non-Invasive Break-Up Time Short-Term Changes with a Water Gradient Silicone Hydrogel Contact Lens. *Life* **2022**, *12*, 1710. https://doi.org/10.3390/life12111710

Academic Editor: Marta Sacchetti

Received: 22 August 2022
Accepted: 24 October 2022
Published: 26 October 2022

1. Introduction

In recent years, soft contact lenses (SCLs), and particularly silicone hydrogels (SHs), have experienced constant changes by the specialized industry, promoting the development of materials, designs and treatments with greater biocompatibility with human tissue, which provide better properties for corneal physiology, eye comfort and wettability [1–4]. Despite the introduction of new materials and surfactants in contact lens design [5,6], SH-SCL users continue to report dryness and eye discomfort at some time in the day [7], representing one of the main causes of leaving CLs [8]. Crucial comfort factors could be related to changes generated by interactions of the tear film within ocular tissues [9].

Although the exact etiology remains unknown, there are numerous factors related to discomfort in SH-SCL users. Some factors may be susceptible to their environment or due to multifactorial circumstances, such as mode of use, materials, wettability, fitting, tear film fluctuations, multipurpose solution composition, hygiene protocol, environmental exposure or patient lifestyle [6,10,11]. Similarly, osmolarity, temperature and even digital devices could impact certain CL parameters, such as thickness, diameter, optical zone and wettability [10,12].

The increase in water content and the combination of surface treatments by the SCL industry has led to an advance in wettability, fulfilling the purpose of reducing contact angle hysteresis, generating greater comfort patterns in CL daily use [13]. The inclusion of moisturizers in the CL matrix or surface suggests an increase in tear film volume and stability [14–16], playing a vital role in comfort [11], which is not the only clinically significant factor [17]. The lacrimal film promotes corneal function, lubricating and protecting the ocular surface [18]. Inserting a contact lens causes an alteration in the tear film dynamics and divides it into two interfaces, the outermost or pre-lens phase and the post-lens internal phase [19]. The study of tear film with different noninvasive techniques, such as tear rupture times (NIBUT) [20], dehydration of pre-SCL film directly carried in the eye (NIDUT) [10], lipid layer interferometry color pattern [21] and tear meniscus height measurement [22], supposes a useful guide to predict changes in tear stability and is a factor of certainty about the success of the adaptation of CLs. Tear film stability is a good indicator of healthy eye function [23]. The newest development thus far is the water gradient SH-SCL, whose dual structure features a 33% water core and continues to progress to the outside with more surface structure of approximately 80% water, presumably designed to improve use tolerance and minimize the problems associated with SH-SCL [24].

The purpose of this study was to evaluate pre-lens tear film volume, stability and lipid interferometry patterns with a silicone hydrogel water content contact lens through a novel, noninvasive, ocular-surface-analyzer technology.

2. Materials and Methods

2.1. Design

This longitudinal, single-center prospective study was conducted at the optometry cabinets in the Pharmacy School of the University of Seville. This research was performed according to the Helsinki Declaration and the Ethical Committee Board of the University of Seville (0384-N-22).

2.2. Subjects

All subjects included in the study read and signed the informed consent form. An informative sheet was provided to all subjects with the detailed study procedure. The inclusion criteria were as follows: (1) healthy subjects without any eye disease or eye treatment, (2) age between 18 and 35 years old, (3) Contact Lens Dry Eye Questionnaire 8 (CLDEQ8) under 12 score points [25], (4) daily or monthly replacement silicone hydrogel contact lens wearers, (5) manifest objective and subjective spherical equivalent refraction $\leq$ 4.50 diopters, and (6) manifest objective and subjective refractive astigmatism $\leq$ 1.00 diopter. The exclusion criteria were as follows: (1) ocular infection or inflammation, with no previous history of ocular surgery, (2) taking any ophthalmic or systemic medications with tear film or ocular surface effects, and (3) pregnancy or breastfeeding.

2.3. Materials

Noninvasive analysis of the tear film was assessed with the Integrated Clinical Platform (ICP) Ocular Surface Analyzer (OSA) from SBM System® (Orbassano, Torino, Italy). Detailed information of the device was described in previous research [26]. Meibomian gland evaluation was assessed with the nonmydriatic infrared meibography digital fundus camera Cobra® HD (Construzione Strumenti Oftalmici CSO®, Firenze, Italy). The degree of meibomian gland dysfunction (MGD) was measured by the ImageJ method defined by

Pult and Nichols [27]. MGD was classified into one of four grades according to the severity of the loss.

Tear volume was measured with Schirmer strips (Tear Flo, HUB Pharmaceutical, Michigan, USA). Two subjective dry eye disease questionnaires were used: the Contact Lens Dry Eye Questionnaire 8 (CLDEQ-8) [25] and the Standard Patient Evaluation of Eye Dryness (SPEED) [28] test.

Regarding the contact lens studied, silicone hydrogel (TOTAL 30®, Alcon Inc., Fort Worth, Texas, USA) was used. The Food and Drug Administration (FDA) material group has a high water content and is nonionic (V-B). This contact lens has biomimetic CELLIGENT® Technology that supports resistance to bacteria and lipid deposits. Furthermore, other features were the water gradient technology within a high water content (>90%) at the outermost surface. The technical parameters are presented in Table 1. The contact lens care system solution was a multipurpose solution (MPS) containing 0.00015% polyhexamethylene biguanide (PHMB), 0.01% ethylenediaminetetraacetic acid (EDTA), sodium hyaluronate and hydroxyethyl cellulose in an isotonic, buffered aqueous solution (Lens 55® Care Hyaluropolimer Plus 360 mL, Servilens Fit and Cover®, Granada, Spain) for all subjects. Lehfilcon A silicone hydrogel technical parameters are presented in Table 1.

Table 1. Lehfilcon A silicone hydrogel technical parameters.

Material	Lehfilcon A
Base Curve	8.4 mm
Diameter	14.2 mm
FDA Group	V-B
Wetting Agent	Phosphoryl Choline
Material/Water (%)	45/55
Center Thickness	0.08 mm
Oxygen Transmission	154 Dk/t
Modulus	0.6 MPa
UV Blocking	Class 1
UVA Blocking	>90%
UVB Blocking	>99%
Light Filter	HEVL
Dynamic Light	No absorption

HEVL: High Energy Visible Light.

2.4. Examination Procedure

In the first phase, subjects were classified according to inclusion and exclusion criteria. The subject's sample was obtained from the non-optometry academic community. Standard contact lens protocol adaptation was performed according to the Graeme Young Soft Lens Design and Fitting chapter in the Nathan Efron *Contact Lens Practice* book [29]. All subjects were trained to prevent using any lubricants or contact lenses seven days prior to the study. After this wash-out period was finished, subjective questionnaires and noninvasive examination with OSA and meibography was performed [26]. Conjunctival redness classification, lipid layer thickness (LLT), tear meniscus height (TMH), first NIBUT (FNIBUT), mean NIBUT (MNIBUT) and lid opening time (LOT) were included in the protocol.

In a second phase, the subjects were reassessed after 30 min of contact lens wearing to quantify the volume, stability and lipid pattern of the short-term pre-lens tear film. The temperature and humidity area assessment conditions were constant during all measurements. Ocular surface tests were taken alternating between both eyes. Furthermore, between OSA measurement steps, the subjects blinked normally within one minute, and prior to the next measurement, the subject deliberately blinked three full times.

2.5. Statistical Analysis

Statistical analysis was performed with SPSS statistical software (version 26.0, IBM Corp, Armonk, NY, USA). Descriptive analysis was performed with the mean ± SD (range value). The normality distribution of the data was assessed with the Shapiro–Wilk test Differences in qualitative variables were assessed with the chi-squared test. The differences between the previous and short-term pre-lens variables were performed with the Wilcoxon test. The correlation study was evaluated with the Spearman's rho test. For all tests, the significance level was established at 95% (p value < 0.05). The sample size was evaluated with the GRANMO® calculator (Institut Municipal d'Investigació Mèdica, Barcelona, Spain. Version 7.12). The two-sided test was used. The risk of alpha and beta was set at 5% and 20%, respectively. The estimated standard deviation (SD) of the differences was set at 0.45 (based on Marx et al. [30] SD main variable research), the expected minimum pre-lens NIBUT difference was set at 0.30 s, and finally, the loss to follow-up rate was set at 0.00. This achieved a recommended sample size of twenty subjects.

3. Results

Sixty-two silicone hydrogel contact fittings were performed in a sample of thirty-one myopic with low astigmatism subjects. Descriptive analyses of sex, nationality, age, noncycloplegic manifest refraction, LogMAR and decimal visual acuity, corneal meridian, contact lens power, Schirmer test, CLDEQ-8 questionnaire, SPEED questionnaire, and superior and inferior eyelid meibomian gland dysfunction are presented in Table 2. Longitudinal ocular surface measurements are presented in Table 3.

Conjunctival redness classification achieved a non-statistically significant increase of 0.06 ± 0.30 grades on the Efron Scale (W = 17.50, p = 0.10). Conjunctival redness decreased, increased and did not change in 5, 1 and 56 eyes, respectively. A trivial effect size of 0.11 was reported. The rho of Spearman between the previous and posterior conjunctival redness classifications was 0.89 (p < 0.01). Lipid layer thickness interferometry decreased 0.14 ± 1.00 grades on the Guillon scale (W = 311.00, p = 0.23). Lipid thickness decreased, increased and did not change in 22, 17 and 23 eyes, respectively. A trivial effect size of 0.09 was reported. The rho of Spearman between the previous and posterior conjunctival redness classifications was 0.72 (p < 0.01). Lipid layer thickness interferometry decreased from grade 2 to grade 0, as presented in Figure 1. The tear meniscus height remained remarkably similar, with a change of 0.001 ± 0.03 mm (W = 695.00, p = 0.76). TMH decreased, increased and did not change in 31, 20 and 11 eyes, respectively. A trivial effect size of 0.01 was reported. The Spearman's rho between the previous and posterior TMH was 0.79 (p < 0.01).

Table 2. Descriptive analysis of the sample.

Variable		Value
Gender (%)		
	Male	14 (22.6)
	Female	48 (77.4)
Nationality (%)		
	Italian	21 (67.75)
	Spanish	4 (12.90)
	Mexican	2 (6.46)
	Slovak	1 (3.22)
	Polish	1 (3.22)
	Germany	1 (3.22)
	Austrian	1 (3.22)
	Age (Years)	22.23 ± 1.39 (19 to 25)

Table 2. *Cont.*

Variable	Value
Sphere (Diopters)	−2.64 ± 1.15 (−5.50 to −0.50)
Cylinder (Diopters)	−0.44 ± 0.37 (−1.50 to 0.00)
Axis (Degrees, °)	111.44 ± 70.08 (5.00 to 180.00)
Visual Acuity (Log MAR)	−0.03 ± 0.05 (−0.10 to 0.10)
Visual Acuity (Decimal)	1.07 ± 0.10 (0.80 to 1.20)
Flat Corneal Meridian (mm)	7.87 ± 0.31 (7.40 to 8.74)
Steep Corneal Meridian (mm)	7.73 ± 0.29 (7.25 to 8.61)
Mean Corneal Meridian (mm)	7.80 ± 0.30 (7.37 to 8.67)
Contact Lens Power (Diopters)	−2.56 ± 1.12 (−5.00 to −0.75)
Schirmer Test (mm)	30.21 ± 8.43 (6.00 to 35.00)
CLDEQ8 (Score Points)	11.32 ± 5.56 (1.00 to 29.00)
SPEED Test (Score Points)	7.39 ± 4.39 (0.00 to 15.00)
Superior Eyelid MGD (%)	28.87 ± 15.11 (10.30 to 96.20)
Inferior Eyelid MGD (%)	49.69 ± 17.86 (17.00 to 87.30)

CLDEQ8: Contact Lens Dry Eye Questionnaire, SPEED: Standard Patient Evaluation of Eye Dryness.

Table 3. Ocular surface longitudinal changes before and with silicone hydrogel wearing.

Variable	Before Lehfilcon A	30-min with Lehfilcon A	*p* Value
Conjunctival Redness Classification (Efron Scale)	1.08 ± 0.63 (0.00 to 2.00)	1.15 ± 0.56 (0.00 to 2.00)	0.10
Lipid Layer Thickness Interferometry (Guillon Pattern)	2.05 ± 1.53 (0.00 to 5.00)	1.90 ± 1.73 (0.00 to 5.00)	0.23
Tear Meniscus Height (Millimeters)	0.21 ± 0.04 (0.11 to 0.32)	0.21 ± 0.06 (0.07 to 0.32)	0.76
First NIBUT (Seconds)	5.03 ± 1.04 (3.60 to 7.80)	4.63 ± 0.89 (3.64 to 8.52)	0.01 *
Mean NIBUT (Seconds)	15.19 ± 9.54 (4.50 to 49.76)	21.27 ± 11.97 (5.44 to 56.48)	<0.01 *
Lid Opening Time (Seconds)	26.36 ± 19.72 (5.04 to 93.60)	38.58 ± 21.78 (7.04 to 107.04)	<0.01 *

NIBUT: Non-Invasive Break Up Time. * Statistically significant within Wilcoxon test.

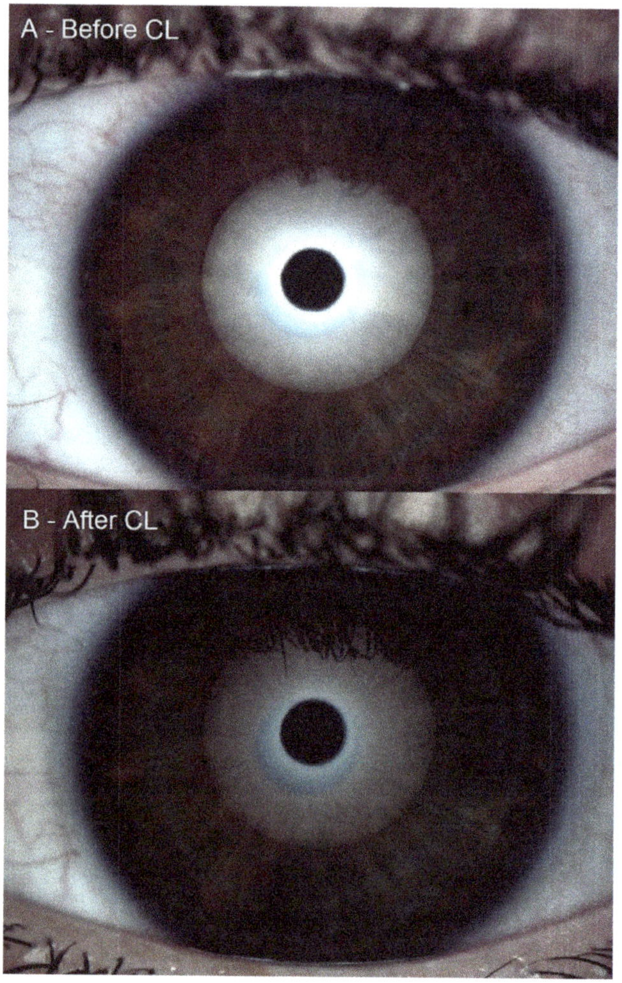

Figure 1. Lipid layer thickness interferometry decreased from grade 2 to grade 0. (**A**): Grade 2 Guillon pattern and (**B**): Grade 0 Guillon pattern.

FNIBUT reported a slight decrease of 0.40 ± 1.40 s (W = 600.00, p = 0.01). FNIBUT decreased, increased and did not change in 41, 20 and 41 eyes, respectively. A 0.41 moderate effect size was reported. The rho of Spearman between the previous and posterior FNIBUT was 0.00 (p = 0.95). A previous FNIBUT achieved a nonsignificant correlation of 0.008 (p = 0.95) with the 20-min FNIBUT. MNIBUT achieved a statistically and clinically significant difference of 6.08 ± 10.40 s. (W = 1156.50, $p < 0.01$). MNIBUT decreased and increased in 17 and 45 eyes, respectively. A 0.56 large effect size was reported. The rho of Spearman between the previous and posterior MNIBUT was 0.57 ($p < 0.01$). A previous MNIBUT achieved a significant correlation of 0.57 ($p < 0.01$). Finally, LOT showed the largest increase of 12.21 ± 19.24 s. (W = 1551.50, $p < 0.01$). LOT decreased, increased and did not change in 18, 43 and 1 eyes, respectively. A 0.58 large effect size was reported. The Spearman's rho between the previous and posterior LOTs was 0.55 ($p < 0.01$). Differences between baseline and short-term results with Lehfilcon A is presented in Table 3. Sequentially captured examples of the initial moment, FNIBUT, MNIBUT and LOT are presented in Figure 2.

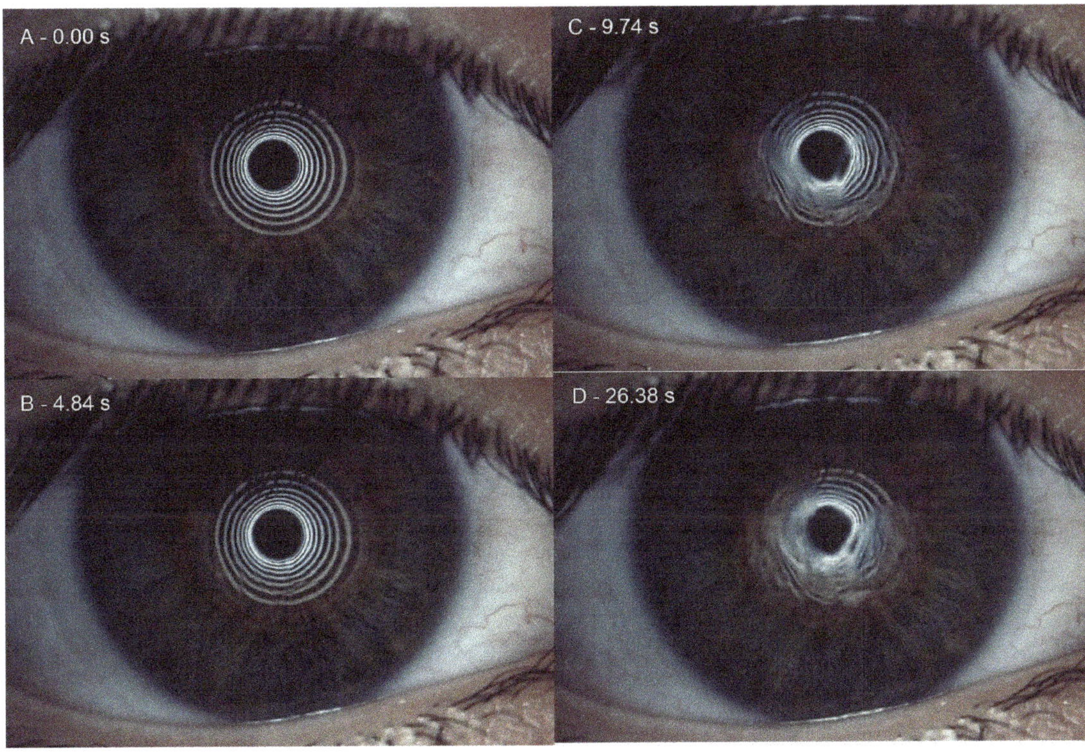

Figure 2. Sequentially captures examples of noninvasive break-up time. (**A**): initial moment, (**B**): first noninvasive break-up time, (**C**): mean noninvasive break-up time and (**D**): last capture prior to lid opening time.

4. Discussion

In this study, the tear film volume, stability and lipid pattern changes within the pre-lens tear film were assessed in a Lehfilcon A silicone hydrogel contact lens with water gradient technology. Moreover, a novel, noninvasive, ocular-surface-analyzer technology was used. The conjunctival redness classification achieved a non-statistically significant increased percentage. Lipid layer thickness interferometry decreased, and tear meniscus height remained remarkably similar to baseline. However, significant statistical and clinical changes were achieved in FNIBUT and MNIBUT that decreased and increased, respectively. Finally, the LOT increased significantly with Lehfilcon A.

These results are similar to those found by Llorens-Quintana et al. [31], who describe how the FNIBUT decreases with CL use time, without finding a relationship between it and the precorneal NIBUT. It also concludes that the changes in the pre-lens NIBUT would be related to the CL material and not only to the quality of the baseline chronic tear film. In a similar line of research, Montani et al. [32] linked these changes to the CL material. The lens with the highest water content and lowest DK (hydrogel) of those studied had less impact on the tear film characteristics, with fewer changes in TMH and pre-lens NIBUT than other lenses assessed with lower water content and higher DK. However, the high water content suggested that the high water gradient contributes to a lower impact on the wearer's tear film, as do other surface treatments [33]. The increase in water content and the combination of surface treatments by CL manufacturers has led to an advance in wettability, generating greater comfort patterns in CL daily use [3,11]. The pre-lens tear film stability could change depending on the wettability of the CL material [34].

The decrease in FNIBUT would be related to the decrease in LLT, which, although it does not present significant changes, has a value lower than that measured before putting in the CL, and that would make the tear evaporate faster. However, the Lehfilcon A aqueous gradient would explain the significant increase in MNIBUT and LOT. Fujimoto et al. [24] reported that daily disposable Lehfilcon A contact lenses increase NIBUT and reduce TMH. The results achieved in our study demonstrated that TMH remains stable in a short-term period, so the integrity of ocular physiology could not vary [35]. Several studies have described the reliability of the pre-lens NIBUT measurement [30,34,36] and the importance of tear film stability to guarantee CL comfort, which, as described by Guillon et al. [37], is lower in patients who present symptoms with CLs than in those who do not. Furthermore, the use of interferometry would be more reliable when obtaining this measurement compared to other methods [34]. Finally, Muhafiz and Demir [38] considered that precorneal NIBUT measurements may be useful for diagnosing tear instability, but that pre-lental NIBUT values are not yet capable of adequately defining tear film dynamics in CL users. We consider that this measurement provides a great deal of information on the relationship between the CL and the tear film and that it can help contrast signs and symptoms.

With respect to limitations, more studies would be necessary to establish a relationship between changes in the lipid layer thickness and the decrease in FNIBUT, as well as the increase in the MNIBUT and LOT. Future research should include the influence of unconventional materials and surface treatment on these parameters to help us choose the appropriate CL for each case, especially in users who already have problems with the tear film.

5. Conclusions

Silicone hydrogel contact lenses with water gradient technology significantly increased the mean pre-lens NIBUT and lid opening time. Lehfilcon A suggested an improvement in contact lens wearers with tear film instability or decreased subjective symptoms of dry eye disease.

Author Contributions: Conceptualization, R.C.-P., M.-J.B.-L., C.M. and J.-M.S.-G. methodology, R.C.-P., M.-J.B.-L., C.M. and J.-M.S.-G.; validation, R.C.-P., M.-J.B.-L., C.M. and J.-M.S.-G.; formal analysis, R.C.-P., M.-J.B.-L., C.M. and J.-M.S.-G.; investigation, R.C.-P., M.-J.B.-L., C.M. and J.-M.S.-G.; resources, R.C.-P., M.-J.B.-L., C.M. and J.-M.S.-G.; data curation, R.C.-P., M.-J.B.-L., C.M. and J.-M.S.-G.; writing—original draft preparation, R.C.-P., M.-J.B.-L., C.M. and J.-M.S.-G.; writing—review and editing, R.C.-P., M.-J.B.-L., C.M. and J.-M.S.-G.; visualization, R.C.-P., M.-J.B.-L., C.M. and J.-M.S.-G.; supervision, R.C.-P., M.-J.B.-L., C.M. and J.-M.S.-G. All authors have read and agreed to the published version of the manuscript.

Funding: This research received no external funding.

Institutional Review Board Statement: The study was conducted in accordance with the Declaration of Helsinki and approved by the Ethics Committee of the University of Seville (0384-N-22).

Informed Consent Statement: Informed consent was obtained from all subjects involved in the study.

Data Availability Statement: The data presented in this study are available on request from the corresponding author. The data are not publicly available due to their containing information that could compromise the privacy of research participants.

Acknowledgments: The authors appreciate the support offered by the members of the Department of Physics of Condensed Matter, Faculty of Physics, University of Seville, with special thanks to Javier Romero-Landa and Clara Conde-Amiano. In addition, the authors also appreciate the technical support offered by the members and facilities of the Faculty of Pharmacy, University of Seville, with special thanks to María Álvarez-de-Sotomayor.

Conflicts of Interest: The authors declare no conflict of interest.

Abbreviations

CL	Contact Lens
CLDEQ8	Contact Lens Dry Eye Questionnaire 8
EDTA	Ethylenediaminetetraacetic Acid
FDA	Food and Drug Administration
FNIBUT	First Non-Invasive Break-Up Time
ICP	Integrated Clinical Platform
LLT	Lipid Layer Thickness
LOT	Lid Opening Time
MGD	Meibomian Gland Dysfunction
MNIBUT	Mean Non-Invasive Break-Up Time
NIBUT	Non-Invasive Break-Up Time
NIDUT	Non-Invasive Dehydration-Up Time
OSA	Ocular Surface Analyzer
SCL	Soft Contact Lens
SH-SCL	Silicone Hydrogel Soft Contact Lens
SPEED	Standard Patient Evaluation of Eye Dryness
TMH	Tear Meniscus Height

References

1. Tahhan, N.; Naduvilath, T.J.; Woods, C.; Papas, E. Review of 20 years of soft contact lens wearer ocular physiology data. *Contact Lens Anterior Eye* **2022**, *45*, 101525. [CrossRef]
2. Jacob, J.T. Biocompatibility in the development of silicone-hydrogel lenses. *Eye Contact Lens* **2013**, *39*, 13–19. [CrossRef]
3. Eftimov, P.B.; Yokoi, N.; Peev, N.; Paunski, Y.; Georgiev, G.A. Relationships between the material properties of silicone hydrogels: Desiccation, wettability and lubricity. *J. Biomater. Appl.* **2021**, *35*, 933–946. [CrossRef]
4. Tauste, A.; Ronda, E.; Baste, V.; Bråtveit, M.; Moen, B.E.; Seguí Crespo, M. del M. Ocular surface and tear film status among contact lens wearers and non-wearers who use VDT at work: Comparing three different lens types. *Int. Arch. Occup. Environ. Health* **2018**, *91*, 327–335. [CrossRef]
5. Musgrave, C.S.A.; Fang, F. Contact lens materials: A materials science perspective. *Materials* **2019**, *12*, 261. [CrossRef]
6. Lira, M.; Silva, R. Effect of Lens Care Systems on Silicone Hydrogel Contact Lens Hydrophobicity. *Eye Contact Lens* **2017**, *43*, 89–94. [CrossRef]
7. Markoulli, M.; Kolanu, S. Contact lens wear and dry eyes: Challenges and solutions. *Clin. Optom.* **2017**, *9*, 41–48. [CrossRef]
8. McMonnies, C.W. Could contact lens dryness discomfort symptoms sometimes have a neuropathic basis? *Eye Vis.* **2021**, *8*, 12. [CrossRef]
9. Rex, J.; Knowles, T.; Zhao, X.; Lemp, J.; Maissa, C.; Perry, S.S. Elemental Composition at Silicone Hydrogel Contact Lens Surfaces. *Eye Contact Lens* **2018**, *44*, S221–S226. [CrossRef]
10. Guillon, M.; Patel, T.; Patel, K.; Gupta, R.; Maissa, C.A. Quantification of contact lens wettability after prolonged visual device use under low humidity conditions. *Contact Lens Anterior Eye* **2019**, *42*, 386–391. [CrossRef]
11. Eftimov, P.; Yokoi, N.; Peev, N.; Georgiev, G.A. Impact of air exposure time on the water contact angles of daily disposable silicone hydrogels. *Int. J. Mol. Sci.* **2019**, *20*, 1313. [CrossRef] [PubMed]
12. Lee, S.E.; Kim, S.R.; Park, M. Oxygen permeability of soft contact lenses in different pH, osmolality and buffering solution. *Int. J. Ophthalmol.* **2015**, *8*, 1037–1042. [CrossRef] [PubMed]
13. Tighe, B.J. A decade of silicone hydrogel development: Surface properties, mechanical properties, and ocular compatibility. *Eye Contact Lens* **2013**, *39*, 4–12. [CrossRef]
14. Maulvi, F.A.; Patel, P.J.; Soni, P.D.; Desai, A.R.; Desai, D.T.; Shukla, M.R.; Ranch, K.M.; Shah, S.A.; Shah, D.O. Novel Poly(vinylpyrrolidone)-Coated Silicone Contact Lenses to Improve Tear Volume during Lens Wear: In Vitro and in Vivo Studies. *ACS Omega* **2020**, *5*, 18148–18154. [CrossRef] [PubMed]
15. Singh, A.; Li, P.; Beachley, V.; McDonnell, P.; Elisseeff, J.H. A hyaluronic acid-binding contact lens with enhanced water retention. *Contact Lens Anterior Eye* **2015**, *38*, 79–84. [CrossRef] [PubMed]
16. Chang, W.H.; Liu, P.Y.; Lin, M.H.; Lu, C.J.; Chou, H.Y.; Nian, C.Y.; Jiang, Y.T.; Hsu, Y.H.H. Applications of hyaluronic acid in ophthalmology and contact lenses. *Molecules* **2021**, *26*, 2485. [CrossRef] [PubMed]
17. García-Montero, M.; Rico-del-Viejo, L.; Llorens-Quintana, C.; Lorente-Velázquez, A.; Hernández-Verdejo, J.L.; Madrid-Costa, D. Randomized crossover trial of silicone hydrogel contact lenses. *Contact Lens Anterior Eye* **2019**, *42*, 475–481. [CrossRef]
18. Bai, Y.; Nichols, J.J. Advances in thickness measurements and dynamic visualization of the tear film using non-invasive optical approaches. *Prog. Retin. Eye Res.* **2017**, *58*, 28–44. [CrossRef]
19. Muntz, A.; Subbaraman, L.N.; Sorbara, L.; Jones, L. Tear exchange and contact lenses: A review. *J. Optom.* **2015**, *8*, 2–11. [CrossRef]
20. Graham, A.D.; Lin, M.C. The relationship of pre-corneal to pre-contact lens non-invasive tear breakup time. *PLoS ONE* **2021**, *16*, e0247877. [CrossRef]

21. Bai, Y.; Ngo, W.; Nichols, J.J. Characterization of the thickness of the tear film lipid layer using high resolution microscopy. *Ocul. Surf.* **2019**, *17*, 356–359. [CrossRef] [PubMed]
22. Binotti, W.W.; Bayraktutar, B.; Ozmen, M.C.; Cox, S.M.; Hamrah, P. A Review of Imaging Biomarkers of the Ocular Surface. *Eye Contact Lens* **2020**, *46*, S84–S105. [CrossRef] [PubMed]
23. Willcox, M.D.P.; Argüeso, P.; Georgiev, G.A.; Holopainen, J.M.; Laurie, G.W.; Millar, T.J.; Papas, E.B.; Rolland, J.P.; Schmidt, T.A.; Stahl, U.; et al. TFOS DEWS II Tear Film Report. *Ocul. Surf.* **2017**, *15*, 366–403. [CrossRef] [PubMed]
24. Fujimoto, H.; Ochi, S.; Yamashita, T.; Inoue, Y.; Kiryu, J. Role of the water gradient structure in inhibiting thin aqueous layer break in silicone hydrogel-soft contact lens. *Transl. Vis. Sci. Technol.* **2021**, *10*, 5. [CrossRef] [PubMed]
25. Chalmers, R.L.; Keay, L.; Hickson-Curran, S.B.; Gleason, W.J. Cutoff score and responsiveness of the 8-item Contact Lens Dry Eye Questionnaire (CLDEQ-8) in a Large daily disposable contact lens registry. *Contact Lens Anterior Eye* **2016**, *39*, 342–352. [CrossRef] [PubMed]
26. Sánchez-González, M.C.; Capote-Puente, R.; García-Romera, M.-C.; De-Hita-Cantalejo, C.; Bautista-Llamas, M.-J.; Silva-Viguera, C.; Sánchez-González, J.-M. Dry eye disease and tear film assessment through a novel non-invasive ocular surface analyzer: The OSA protocol. *Front. Med.* **2022**, *9*, 938484. [CrossRef]
27. Pult, H.; Nichols, J.J. A review of meibography. *Optom. Vis. Sci.* **2012**, *89*, E760–E769. [CrossRef]
28. Baudouin, C.; Aragona, P.; Van Setten, G.; Rolando, M.; Irkeç, M.; Del Castillo, J.B.; Geerling, G.; Labetoulle, M.; Bonini, S. Diagnosing the severity of dry eye: A clear and practical algorithm. *Br. J. Ophthalmol.* **2014**, *98*, 1168–1176. [CrossRef] [PubMed]
29. Young, G. Soft Lens Design and Fitting. In *Contact Lens Practice*; Efron, N., Ed.; Elsevier: Amsterdam, The Netherlands, 2018; pp. 86–94.e1. ISBN 978-0-7020-6660-3.
30. Marx, S.; Eckstein, J.; Sickenberger, W. Objective analysis of pre-lens tear film stability of daily disposable contact lenses using ring mire projection. *Clin. Optom.* **2020**, *12*, 203–211. [CrossRef]
31. Llorens-Quintana, C.; Mousavi, M.; Szczesna-Iskander, D.; Iskander, D.R. Non-invasive pre-lens tear film assessment with high-speed videokeratoscopy. *Contact Lens Anterior Eye* **2018**, *41*, 18–22. [CrossRef] [PubMed]
32. Montani, G.; Martino, M. Tear film characteristics during wear of daily disposable contact lenses. *Clin. Ophthalmol.* **2020**, *14*, 1521–1531. [CrossRef]
33. Vidal-Rohr, M.; Wolffsohn, J.S.; Davies, L.N.; Cerviño, A. Effect of contact lens surface properties on comfort, tear stability and ocular physiology. *Contact Lens Anterior Eye* **2018**, *41*, 117–121. [CrossRef] [PubMed]
34. Itokawa, T.; Suzuki, T.; Iwashita, H.; Hori, Y. Comparison and evaluation of prelens tear film stability by different noninvasive in vivo methods. *Clin. Ophthalmol.* **2020**, *14*, 4459–4468. [CrossRef] [PubMed]
35. Müller, C.; Marx, S.; Wittekind, J.; Sickenberger, W. Subjective comparison of pre-lens tear film stability of daily disposable contact lenses using ring mire projection. *Clin. Optom.* **2020**, *12*, 17–26. [CrossRef] [PubMed]
36. Marx, S.; Sickenberger, W. A novel in-vitro method for assessing contact lens surface dewetting: Non-invasive keratograph dry-up time (NIK-DUT). *Contact Lens Anterior Eye* **2017**, *40*, 382–388. [CrossRef]
37. Guillon, M.; Dumbleton, K.A.; Theodoratos, P.; Wong, S.; Patel, K.; Banks, G.; Patel, T. Association between contact lens discomfort and pre-lens tear film kinetics. *Optom. Vis. Sci.* **2016**, *93*, 881–891. [CrossRef]
38. Muhafiz, E.; Demir, M.S. Ability of non-invasive tear break-up time to determine tear instability in contact lens wearers. *Int. Ophthalmol.* **2022**, *42*, 959–968. [CrossRef] [PubMed]

life

MDPI

Article

Prevalence of Dry Eyes Symptoms in Association with Contact Lenses and Refractive Status in Portugal

Miguel Ángel Sánchez-Tena [1,2], Clara Martinez-Perez [2,*], Cristina Alvarez-Peregrina [1] and Núcleo de Investigação Aplicada em Ótica e Optometria [2,†]

1 Department of Optometry and Vision, Faculty of Optics and Optometry, Universidad Complutense de Madrid, 28040 Madrid, Spain
2 ISEC LISBOA-Instituto Superior de Educação e Ciências, 1750-179 Lisboa, Portugal
* Correspondence: clara.perez@iseclisboa.pt
† Núcleo de Investigação Aplicada em Ótica e Optometria are listed in Acknowledgments.

Simple Summary: The objective of this study is to determine the presence of ocular symptoms in soft-contact-lens wearers that change according to refractive status. To do so, the CLDEQ-8 questionnaire was administered during the months of January to March 2022. Significant differences have been found based on the symptoms present with contact lenses and the degree of myopia. The intensity of visual disturbances was higher in the participants with medium myopia compared to those with low and high myopia. In conclusion, contact-lens users with hyperopia showed a higher rate of ocular dryness than those with myopia. In turn, wearing daily-replacement lenses could be one of the reasons for the lesser presence of ocular dryness when compared to monthly-replacement lenses.

Abstract: Background: Determine whether the presence of ocular symptoms in soft-contact-lens wearers changes depending on the refractive status. Methods: During the months of January to March 2022, the CLDEQ-8 questionnaire was administered to soft-contact-lens wearers. The statistical analysis was carried out using the SPSS 27.0 computer program (SPSS Inc., Chicago, IL, USA). Results: A total of 251 subjects participated in the study, with a higher percentage of myopes than hyperopes (82.1% versus 16.7%; $p < 0.001$). Out of all total participants, 21.5% suffered from dry-eye symptoms. It was noted that hyperopes presented a higher rate of dry-eye symptoms ($p = 0.041$). At the same time, the spherical equivalent was more positive in the participants with dry-eye symptoms ($p = 0.014$). Significant differences were found based on the symptoms present with contact lenses and the degree of myopia. The intensity of visual disturbances was higher in the participants with medium myopia (median [IQR]: 1/5 [2]) compared to those with low (median [IQR]: 0/5 [2]) and high myopia (median [IQR]: 0/5 [1]) ($p = 0.009$). Conclusions: Contact-lens wearers with hyperopia showed a higher rate of ocular dryness than those with myopia. In turn, wearing daily-replacement lenses could be one of the reasons for the lesser presence of ocular dryness compared to monthly-replacement lenses.

Keywords: ocular discomfort; CLDEQ-8; dry eye; refractive errors

Citation: Sánchez-Tena, M.Á.; Martinez-Perez, C.; Alvarez-Peregrina, C.; Núcleo de Investigação Aplicada em Ótica e Optometria. Prevalence of Dry Eyes Symptoms in Association with Contact Lenses and Refractive Status in Portugal. *Life* **2022**, *12*, 1656. https://doi.org/10.3390/life12101656

Academic Editors: José-María Sánchez-González, Carlos Rocha-de-Lossada and Alejandro Cerviño

Received: 20 September 2022
Accepted: 19 October 2022
Published: 20 October 2022

1. Introduction

Around 140 million people throughout the world wear contact lenses (CL) to correct their refractive errors [1]. Despite the advances that have been made in terms of CL technology, this number has remained stable over the last decade. The main reason for this stability is that 10% to 50% of users stop wearing their CL after 3 years due to discomfort. In fact, 70% of CL wearers report that they experience discomfort by the end of the day, and 40% of soft-CL wearers state that they experience dry eye, 25% of whom report moderate to severe symptoms, resulting in a reduction in the time wearing CL [2–7].

In 2013, the Tear Film and Ocular Surface Society (TFOS) defined CL discomfort as *"a condition characterized by episodic or persistent adverse ocular sensations related contact lens* wear", which is caused by a "*reduced compatibility between the contact lenses and the ocular environment*" [8].

Nowadays, soft CL made from silicone hydrogel are used to correct different types of ametropia. These CL boast a higher oxygen-transmission coefficient than conventional soft CL. The hydrophilic material boasts a good surface-moisturizing capacity, which is sufficient for both tearing and transporting fluids across the CL. Thanks to the specific properties offered, silicone-hydrogel CL offer the best oxygen-transmission rate, therefore making them the most adequate option from a physiological standpoint [9]. However, silicone increases the elasticity of the CL, and some users struggle to tolerate this. In addition, the high oxygen transmissibility activates the peroxidation of proteins in the cornea. Dehydration is one of the main causes of dryness and discomfort. Considering their structure, silicone-hydrogel CL have a high moisture content [9], and the higher water content in the CL results in faster evaporation, with dehydrated conventional soft CL becoming absorbent of the water contained in the tear film. Silicone-hydrogel CL are related to those artificial factors, which lead to a decrease in the tear-film stability in the presence of several factors [10].

Current research on new advances in silicone-hydrogel contact-lens polymers and lens-care products focuses primarily on how to minimize the impact on eye health and increase comfort. The goal is to improve the CL-wearing experience [11,12]. Thus, the new products mainly consider the interaction of the lens with the ocular surface in order to minimize the mechanical and physiological effects on the eye.

Nevertheless, despite all these advances, discomfort caused by dry eye remains the main reason why people choose to discontinue CL wear, and it is the main cause of frustration among both patients and doctors [8].

In 2002, the Contact Lens Dry Eye Questionnaire (CLDEQ) [3,13] was designed and validated to assess dry-eye symptoms among CL wearers. The long version included questions that covered the patient's CL-wearing history, frequency, use, and presence of ocular-surface symptoms, along with questions about treatments, computer use, and environmental factors.

In the year 2009, an abbreviated version of this questionnaire was developed and validated with eight questions (CLDEQ-8) designed to evaluate the severity of dry-eye symptoms in soft-CL users in the past 2 weeks [14,15]. Each question was answered using a Likert scale of 0 to 4, 0 to 5, or 1 to 6, and the total score was evaluated based on a scale from 0 to 37. The cut-off score for dry-eye diagnosis was established at $\geq$12 points [15,16].

The objective of this study was to determine whether the presence of ocular symptoms in soft-CL users varied depending on the refractive status.

2. Materials and Methods

2.1. Study Design and Data Gathering

A prospective, transversal, and observational study was carried out in soft-contact-lens wearers. The CLDEQ-8 questionnaire validated by Chalmers et al. [15] was used as well as questions about demographic, prescription, type, and CL replacement, with the aim of evaluating the severity of dry-eye symptoms in soft-CL wearers in the previous 2 weeks. The cutoff value for dry-eye disease was set to $\geq$12 points. The supplementary material shows the CLDEQ-8 questionnaire that was used [15].

The data were collected from January to March 2022 throughout the entire region of Lisbon, Portugal.

2.2. Statistical Analysis

The statistical analysis was carried out using the SPSS 27.0 computer software (SPSS Inc., Chicago, IL, USA). The normal distribution of the variables was conducted using the Kolmogorov–Smirnov test, establishing a significance level of 0.05. The Wilcoxon rank-sum test was used because of nonparametric distribution. The chi-squared test or Fisher's exact test were used, as appropriate, to check the association between the categorical variables.

Spearman's correlation was used for analyzing the quantitative variables. A cutoff point of $p \leq 0.05$ was considered to assess statistical significance.

To determine the refractive state, the spherical-equivalent formula (SE = SE = sphere + cylinder/2) was used. A patient was considered myopic when the SE was more negative than or equal to −0.50D, hyperopic when it was more positive or equal to +0.50D, and emmetropic when it was between −0.25D and +0.25D [17,18].

The research was conducted in accordance with the principles of the Declaration of Helsinki and with the approval of the ethics committee of the Institute of Education and Science (ISEC Lisbon) on 5 November 2021 recorded under code CE/2022/03/01.

3. Results

3.1. Demographic Data

A total of 248 subjects participated in the study, with a higher percentage of myopes than hyperopes (83.1% versus 16.9%; $p < 0.001$).

The ages of the participants were 35.22 ± 12.95 years and the median and interquartile range was 34 [20]. With regards to gender, the percentage of women was similar to that of men (56.2% versus 43.8%, respectively; $p > 0.05$). Table 1 shows the demographic data and contact-lens information of the analyzed sample.

Table 1. Demographic data and contact-lens information of the study population.

	Total	Hyperopia	Myopia	*p*-Value
No. of participants (% of the total)	248	42 (16.9%)	206 (83.1%)	$p < 0.001$
Gender				
Women	139 (56.0%)	21 (50.0%)	118 (57.3%)	0.399
Men	109 (44.0%)	21 (50.0%)	88 (42.7%)	
Age				
Mean ± SD	35.25 ± 12.96	45.83 ± 14.02	33.10 ± 11.64	$p < 0.001$
Median [IQR]	34.00 [21]	48.50 [18]	32.00 [18]	
SE				
Mean ± SD	−2.65 ± 3.42	2.71 ± 2.43	−3.75 ± 2.42	$p < 0.001$
Median [IQR]	−2.78 [3.15]	2.12 [1.77]	−3.37 [2.95]	
CL Replacement				
Daily	101 (40.7%)	19 (45.2%)	82 (39.8%)	$p = 0.045$
Biweekly	22 (8.7%)	0 (0.0%)	22 (10.7%)	
Monthly	117 (47.2%)	21 (50.0%)	96 (46.6%)	
Quarterly	1 (0.4%)	1 (2.4%)	0 (0.0%)	
Annual	7 (2.8%)	1 (2.4%)	6 (2.9%)	

SD: standard deviation; IRQ: interquartile range; SE: spherical equivalent; CL: contact lens; NA: not available (lack of adequate sample size for statistical analysis).

3.2. CLDEQ-8 Questionnaire

Out of all total participants, 21.4% suffered from dry eye. Thus, the median and interquartile range of patients with healthy eyes was 4/37 [7.00] and that of patients with dry eye was 17/37 [7.25]. Among the myopic patients, 81.6% had healthy eyes (median [IQR]: 3.5/37 [7.00]) and 18.4% suffered from dry eye (median [IQR]: 17/37 [6.50]). Figure 1 shows the score of the myopes in each of the questions.

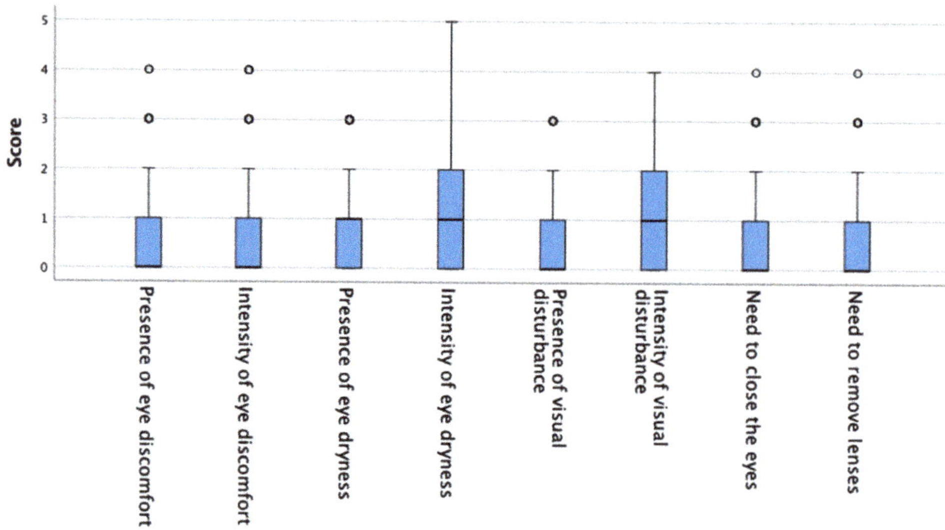

Figure 1. Myopes score on each of the questions of the CLDEQ-8 questionnaire. Box = 1 SD, line = median, whisker = confidence interval 95%, o = extreme values.

Among the hyperopic participants, 42.1% had healthy eyes (median [IQR]: 4/37 [6.00]) and 57.9% suffered from dry eye (median [IQR]: 17/37 [9.00]). Figure 2 shows the score of the hyperopes in each of the questions.

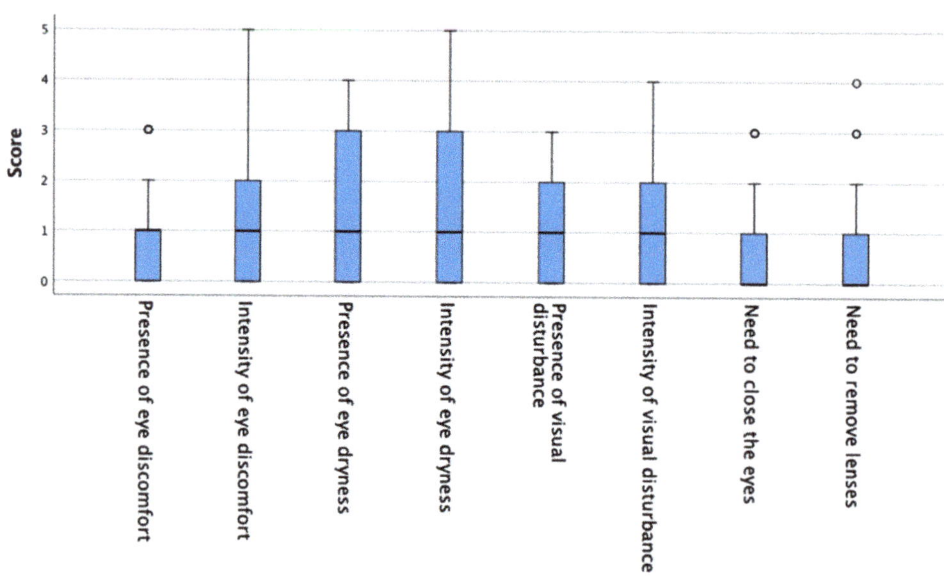

Figure 2. Hyperopes score in each of the questions of CLDEQ-8 questionnaire. Box = 1 SD, line = median, whisker = confidence intervals 95%, o = extreme values.

When analyzing the frequency of contact-lens replacement, it was found that those who wear daily CL presented less sensation of ocular dryness than wearers of monthly and biweekly CL ($p = 0.008$). In turn, the intensity of ocular discomfort was also lower in daily-replacement CL wearers ($p = 0.037$).

When comparing the presence of dry eye based on the refractive state, it was noted that hyperopes presented a higher rate of dry eye ($p = 0.022$). Regarding the presence of other symptoms, no significant differences were found ($p > 0.05$) (Figure 3). Regarding the need to remove the contact lenses, of the total number of participants, 67.7% never needed to, 16.5% less than once a week, 10.1% weekly, 4.8% several times per week, and 0.8% daily. No differences were found based on refractive status, either ($p > 0.05$).

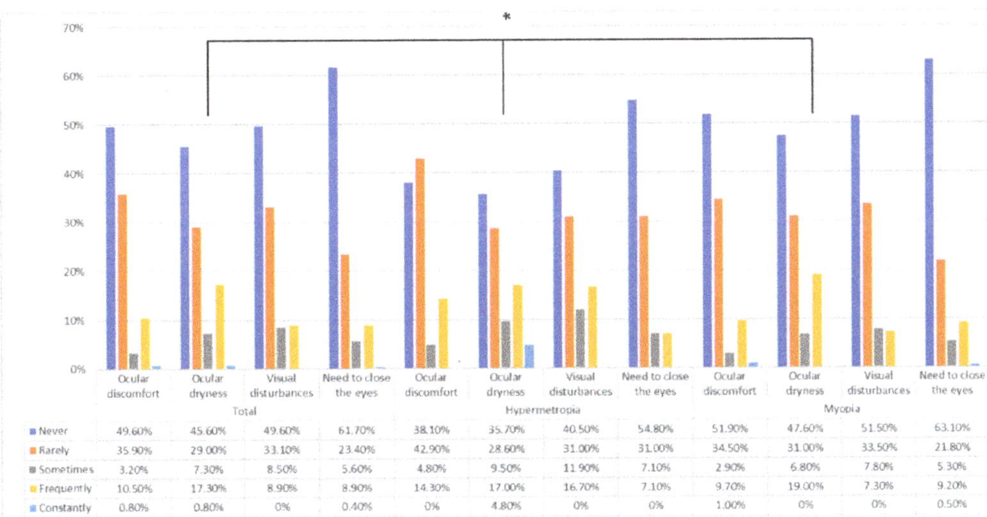

Figure 3. Symptomatology presences in the total number of participants and according to the refractive state. * Significant differences ($p < 0.05$).

At the same time, the spherical equivalent was more positive in the participants with dry eye ($p = 0.014$). Nevertheless, no significant differences were found in the need to remove CL based on the refractive error ($p > 0.05$).

Figure 3 shows the presence of ocular symptoms depending on the refractive state (emmetropic participants were excluded due to the small sample size).

Among the participants with myopia, 43.7% (n = 90) had low myopia, 44.2% had medium myopia (n = 91), and 12.1% had high myopia (n = 25). As shown in Figure 4, significant differences were found based on the symptoms present with contact lenses and the degree of myopia. Thus, the presence of symptoms of ocular discomfort ($p = 0.019$) and symptoms of visual disturbances (0.027) were more frequent with moderate myopia. No significant differences were found in the other symptoms ($p > 0.05$). At the same time, the intensity of visual disturbances was higher in the participants with medium myopia (median [IQR]: 1/5 [2]) compared to those with low (median [IQR]: 0/5 [2]) and high myopia (median [IQR]: 0/5 [1]) ($p = 0.009$). Regarding the need to remove contact lenses, there were no significant differences depending on the degree of myopia ($p > 0.05$).

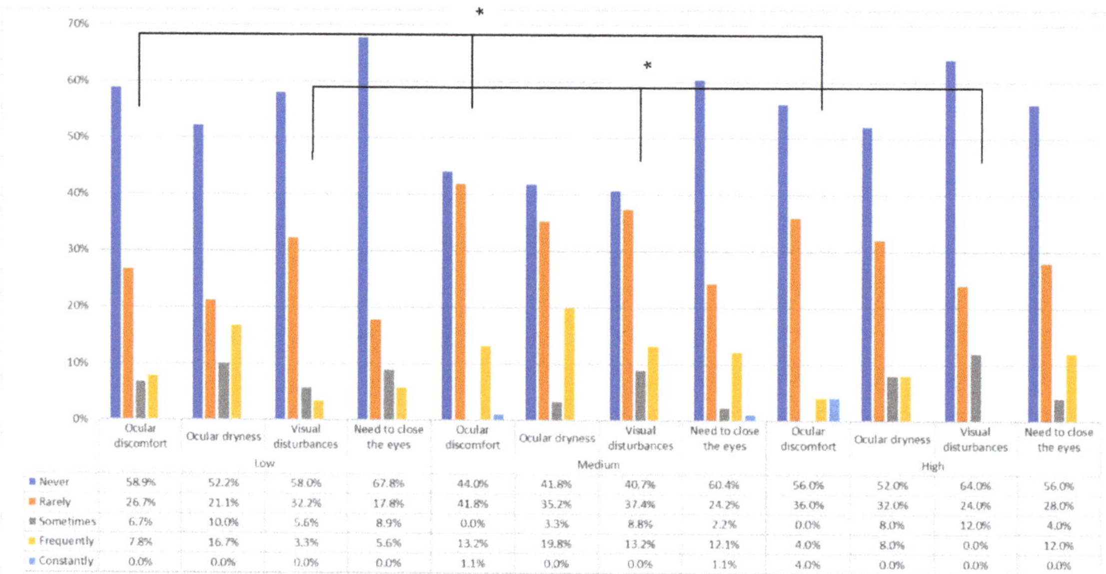

Figure 4. Symptomatology presence according to the degree of myopia. Low: −0.5D < SE > −3D; medium: −3D ≤ SE > −6D; high: SE ≤ −6D. * Significant differences ($p < 0.05$).

4. Discussion

This study showed that the prevalence of dry-eye symptoms in soft-contact-lens wearers in Portugal was 21.4%. This percentage is similar to the results of the study carried out by Vidotti et al. [19], which recorded a prevalence of 27.4% among medical students in Brazil. The results in this study were much higher than those recorded in the study by Abbouda et al. [20] in which a prevalence among Italian teenagers of 9% was recorded, and much lower than those recorded in the studies conducted by García León et al. [21] and Uchino et al. [22], which had a prevalence of 93.9% and 36.1% in university and high school students, respectively. This difference could be explained by the prevalence of refractive errors in the different populations [23,24] or the higher prevalence of high refractive errors in East Asia [25].

Regarding the frequency with which CLs are replaced, in our study we found that monthly and daily CL were the most used. This agrees with the studies by Mohidin et al. [26] and Garcia León et al. [21] in which the most used type of LC was monthly. The high percentage of wearers with daily CL could be due to the great improvements that have been introduced in recent years in terms of materials to ensure better visual quality and improve well-being [27].

In this study, 49.6% of LC wearers declared that they did not experience symptoms of ocular discomfort and 35.9% rarely experienced them. This is in line with the results of the study by Papas et al. [28], in which participants stated that they did not experience symptoms of discomfort within the first 8 h of CL usage. One of the reasons why in our study there were hardly any symptoms of ocular discomfort may be associated with the fact that the levels of discomfort increase when the CL is in contact with the eye for a prolonged period, as the eye seems to "get tired" of wearing the CL, with increased discomfort reported after a prolonged period of use. Given that 40.7% of the participants in our study worn daily CL and that the amount of lens care required was minimal, this could explain why few participants reported symptoms of eye discomfort.

It is worth noting that in our study, as with the symptoms of ocular discomfort, 72% of participants stated that they did not experience, or "rarely" experienced, symptoms of ocular dryness. This, contrasted with the results of the study by Alamri et al. [29], in which 62.2% of participants claimed that they suffered from symptoms of ocular dryness, is in line with the results of the study by Rahmawaty Lubis et al. [30], in which 64.1% of participants stated that they did not experience or "rarely" experienced symptoms of ocular dryness. As previously mentioned, the absence of symptoms of ocular discomfort may be because daily CL replacement could help reduce the symptoms of dry eyes.

At the same time, in our study it was found that 5.6% and 3.3% of the subjects presented occasional and frequent symptoms of visual disturbances, respectively. These results are similar to those of the study by Rahmawaty Lubis et al. [30] and far lower than those recorded in the studies by Mohamad Daud et al. [31] and Sapkota et al. [32]. However, both studies found that these symptoms increased with the use of digital devices; therefore, it could be interesting to consider this point in future studies involving daily-CL wearers to determine the reason for the difference in the symptoms of blurry vision.

On the other hand, 67.8% of the participants in our study did not need to close their eyes or remove the LC. Similar rates were recorded in the study by Mohamad Daud et al. [31]. This could be because our study recorded good compliance in terms of the duration of CL wear. In addition, most participants were aged between 30 and 50 years, which is an age range with a lower rate of corneal infiltrates. It is worth noting that in the study carried out by Rahmawaty Lubis et al. [30], the need to close the eyes or remove the CL was higher. This difference could be explained by the fact that this study was conducted in a location in which it rained almost every day of the month, with fresh and moist conditions.

With regards to the difference between the ocular symptoms based on the refractive state, our study observed a higher prevalence of ocular dryness among hyperopic participants. This went against the results recorded in the study by Alamri et al. [29], in which a higher rate of dryness was recorded amongst myopic patients, and our results agree with the study by Fahmy et al. [33], in which hyperopes presented severe dry eye. Until now, the cause of the presence of dry eye according to the refractive state has not been studied. As for myopia, it is suspected that they may have a drier eye, since the lengthened eye can lead to changes in the ocular surface. However, to compare these results accurately, it would be necessary to know the degree of myopia analyzed in both studies to determine the reason for these discrepancies.

One of the limitations of this study was that the wearing time and the number of years wearing CL were not considered, since the CLDEQ-8 questionnaire does not include those questions. On the other hand, our results are based on a validated questionnaire that measure dry-eye symptomatology, and we did not include any clinical measurement to evaluate dry-eye signs. In future studies, it would be interesting to include these measurements.

5. Conclusions

In conclusion, this study found that hyperopic CL wearers experienced a higher level of ocular dryness than myopia participants. In turn, daily-replacement CL wearers had a lower presence of ocular dryness than those with monthly replacement.

Author Contributions: Conceptualization, C.M.-P., C.A.-P. and M.Á.S.-T.; methodology, C.M.-P., C.A.-P. and M.Á.S.-T.; software, C.M.-P., C.A.-P. and M.Á.S.-T.; validation, C.M.-P., C.A.-P. and M.Á.S.-T.; formal analysis, C.M.-P., C.A.-P. and M.Á.S.-T.; investigation, C.M.-P., C.A.-P., M.Á.S.-T. and Núcleo de Investigação Aplicada em Ótica e Optometria; resources, C.M.-P., C.A.-P., M.Á.S.-T. and Núcleo de Investigação Aplicada em Ótica e Optometria; data curation, Núcleo de Investigação Aplicada em Ótica e Optometria; writing—original draft preparation, C.M.-P., C.A.-P. and M.Á.S.-T.; writing—review and editing, C.M.-P., C.A.-P. and M.Á.S.-T.; visualization, C.M.-P., C.A.-P. and M.Á.S.-T.; supervision, C.M.-P., C.A.-P. and M.Á.S.-T.; project administration and funding acquisition, C.A.-P. and M.Á.S.-T. All authors have read and agreed to the published version of the manuscript.

Funding: This research received no external funding.

Institutional Review Board Statement: The study was conducted in accordance with the Declaration of Helsinki and approved by the or Ethics Committee of the Institute of Education and Science (ISEC Lisbon) on 5 November 2021 recorded under code CE/2022/03/01.

Informed Consent Statement: Informed consent was obtained from all subjects involved in the study.

Data Availability Statement: Not applicable.

Acknowledgments: The Núcleo de Investigação Aplicada em Ótica e Optometria is made up of the following members: Ana Cristina Bastos da Silva, Raquel Blanco Cotovio, Carlos Manuel Fernández da Silva, André Pedro Jorge Policarpo, and Carlos Jorge Nunes Cruz Tavares.

Conflicts of Interest: The authors declare no conflict of interest.

References

1. Stapleton, F.; Keay, L.; Jalbert, I.; Cole, N. The epidemiology of contact lens related infiltrates. *Optom. Vis. Sci.* **2007**, *84*, 257–272. [CrossRef] [PubMed]
2. Pritchard, N.; Fonn, D.; Brazeau, D. Discontinuation of contact lens wear: A survey. *Int. Contact Lens. Clin.* **1999**, *26*, 157–162. [CrossRef]
3. Begley, C.G.; Chalmers, R.L.; Mitchell, G.L.; Nichols, K.K.; Caffery, B.; Simpson, T.; DuToit, R.; Portello, J.; Davis, L. Characterization of ocular surface symptoms from optometric practices in North America. *Cornea* **2001**, *20*, 610–618. [CrossRef] [PubMed]
4. Riley, C.; Young, G.; Chalmers, R. Prevalence of ocular surface symptoms, signs, and uncomfortable hours of wear in contact lens wearers: The effect of refitting with daily-wear silicone hydrogel lenses (Senofilcon a). *Eye Contact Lens* **2006**, *32*, 281–286. [CrossRef]
5. Kastelan, S.; Lukenda, A.; Salopek-Rabatić, J.; Pavan, J.; Gotovac, M. Dry eye symptoms and signs in long-term contact lens wearers. *Coll. Antropol.* **2013**, *37*, 199–203.
6. Reddy, S.C.; Ying, K.H.; Theng, L.H.; How, O.T.; Fu-Xiang, K.; bin Mohamed Sikander, M.M. A survey of dry eye symptoms in contact lens wearers and non-contact lens wearers among university students in Malaysia. *J. Clin. Exp. Ophthalmol.* **2016**, *7*, 522. [CrossRef]
7. Chalmers, R.L.; Young, G.; Kern, J.; Napier, L.; Hunt, C. Soft Contact Lens-Related Symptoms in North America and the United Kingdom. *Optom. Vis. Sci.* **2016**, *93*, 836–847. [CrossRef]
8. Nichols, K.K.; Redfern, R.L.; Jacob, J.T.; Nelson, J.D.; Fonn, D.; Forstot, S.L.; Huang, J.F.; Holden, B.A.; Nichols, J.J.; members of the TFOS International Workshop on Contact Lens Discomfort. The TFOS International Workshop on Contact Lens Discomfort: Report of the definition and classification subcommittee. *Investig. Ophthalmol. Vis. Sci.* **2013**, *54*, TFOS14–TFOS19. [CrossRef]
9. Sheedy, J.; Hardy, R.F.; Hayes, J.R. Progressive addition lenses–measurements and ratings. *Optometry* **2006**, *77*, 23–39. [CrossRef]
10. Carnt, N.; Stapleton, F. Strategies for the prevention of contact lens-related Acanthamoeba keratitis: A review. *Ophthalmic Physiol. Opt.* **2016**, *36*, 77–92. [CrossRef]
11. Craig, J.P.; Willcox, M.D.; Argüeso, P.; Maissa, C.; Stahl, U.; Tomlinson, A.; Wang, J.; Yokoi, N.; Stapleton, F.; members of TFOS International Workshop on Contact Lens Discomfort. The TFOS International Workshop on Contact Lens Discomfort: Report of the contact lens interactions with the tear film subcommittee. *Investig. Ophthalmol. Vis. Sci.* **2013**, *54*, TFOS123–TFOS156. [CrossRef]
12. Efron, N.; Jones, L.; Bron, A.J.; Knop, E.; Arita, R.; Barabino, S.; McDermott, A.M.; Villani, E.; Willcox, M.D.; Markoulli, M.; et al. The TFOS International Workshop on Contact Lens Discomfort: Report of the contact lens interactions with the ocular surface and adnexa subcommittee. *Investig. Ophthalmol. Vis. Sci.* **2013**, *54*, TFOS98–TFOS122. [CrossRef]
13. Begley, C.G.; Caffery, B.; Nichols, K.K.; Chalmers, R. Responses of contact lens wearers to a dry eye survey. *Optom. Vis. Sci.* **2000**, *77*, 40–46. [CrossRef]
14. Chalmers, R.L.; Begley, C.G.; Moody, K.; Hickson-Curran, S.B. Contact Lens Dry Eye Questionnaire-8 (CLDEQ-8) and opinion of contact lens performance. *Optom. Vis. Sci.* **2012**, *89*, 1435–1442. [CrossRef]
15. Chalmers, R.L.; Keay, L.; Hickson-Curran, S.B.; Gleason, W.J. Cutoff score and responsiveness of the 8-item Contact Lens Dry Eye Questionnaire (CLDEQ-8) in a Large daily disposable contact lens registry. *Cont. Lens Anterior Eye* **2016**, *39*, 342–352. [CrossRef]
16. Nichols, J.J.; Mitchell, G.L.; Nichols, K.K.; Chalmers, R.; Begley, C. The performance of the contact lens dry eye questionnaire as a screening survey for contact lens-related dry eye. *Cornea* **2002**, *21*, 469–475. [CrossRef]
17. Flitcroft, D.I.; He, M.; Jonas, J.B.; Jong, M.; Naidoo, K.; Ohno-Matsui, K.; Rahi, J.; Resnikoff, S.; Vitale, S.; Yannuzzi, L. IMI-Defining and Classifying Myopia: A Proposed Set of Standards for Clinical and Epidemiologic Studies. *Investig. Ophthalmol. Vis. Sci.* **2019**, *60*, M20–M30. [CrossRef]
18. Galvis, V.; Tello, A.; Camacho, P.A.; Gómez, L.M.; Rey, J.J.; Serrano, A.A. Definition of refractive errors for research studies: Spherical equivalent could not be enough. *J. Optom.* **2021**, *14*, 224–225. [CrossRef]

19. Vidotti, V.G.; Kamegasawa, A. Perfil dos alunos usuários de lentes de contato do curso de Medicina da Universidade Estadual Paulista-UNESP–Botucatu [Profile of medical students from the Universidade Estadual Paulista-UNESP–Botucatu, who wear contact lenses]. *Arq. Bras. Oftalmol.* **2006**, *69*, 197–201. [CrossRef]
20. Abbouda, A.; Restivo, L.; Bruscolini, A.; Pirraglia, M.P.; De Marco, F.; La Cava, M.; Pivetti Pezzi, P. Contact Lens Care among Teenage Students in Italy: A Cross-Sectional Study. *Semin Ophthalmol.* **2016**, *31*, 226–232. [CrossRef]
21. Garcia Leon, M.; de Fátima Arroyo, L.; Ibañez Garcia, M.; Villarreal Calderon, J.R.; Hernández Morales, X.; Chapa de la Peña, A.; Cárdenas Rodríguez, I.I.; Díaz Gómez, M. Patrón de uso de lentes de contacto y sintomatología asociada en estudiantes universitarios. *Rev. Mex. Oftalmol.* **2017**, *91*, 9–17. [CrossRef]
22. Uchino, M.; Dogru, M.; Uchino, Y.; Fukagawa, K.; Shimmura, S.; Takebayashi, T.; Schaumberg, D.A.; Tsubota, K. Japan Ministry of Health study on prevalence of dry eye disease among Japanese high school students. *Am. J. Ophthalmol.* **2008**, *146*, 925–929.e2. [CrossRef]
23. Chin, M.P.; Siong, K.H.; Chan, K.H.; Do, C.W.; Chan, H.H.; Cheong, A.M. Prevalence of visual impairment and refractive errors among different ethnic groups in schoolchildren in Turpan, China. *Ophthalmic Physiol. Opt.* **2015**, *35*, 263–270. [CrossRef]
24. Silva, J.C.; Mújica, O.J.; Vega, E.; Barcelo, A.; Lansingh, V.C.; McLeod, J.; Limburg, H. A comparative assessment of avoidable blindness and visual impairment in seven Latin American countries: Prevalence, coverage, and inequality. *Rev. Panam. Salud Publica* **2015**, *37*, 13–20.
25. Wu, L.J.; You, Q.S.; Duan, J.L.; Luo, Y.X.; Liu, L.J.; Li, X.; Gao, Q.; Zhu, H.P.; He, Y.; Xu, L.; et al. Prevalence and associated factors of myopia in high-school students in Beijing. *PLoS ONE* **2015**, *10*, e0120764. [CrossRef]
26. Mohidin, N.; Lee Fung, T. A Survey of Optomtris Contact Lens Prescribing in Malaysia. *J. Sains Kesihat. Malays.* **2009**, *7*, 59–72.
27. Sulley, A.; Dumbleton, K. Silicone hydrogel daily disposable benefits: The evidence. *Cont. Lens Anterior Eye* **2020**, *43*, 298–307. [CrossRef]
28. Papas, E.; Tilia, D.; McNally, J.; de la Jara, P.L. Ocular discomfort responses after short periods of contact lens wear. *Optom. Vis. Sci.* **2015**, *92*, 665–670. [CrossRef]
29. Alamri, A.; Amer, K.A.; Aldosari, A.A.; Al-Muhsin, S.D.; Al-Maalwi, R.S.; Al Hamdan, S.A.; Al-Tarish, L.M. Assessment of Dry Eye Syndrome Among Contact Lens Users in Asir Region, Saudi Arabia. *Cureus* **2022**, *14*, e21526. [CrossRef]
30. Lubis, R.R.; Gultom, M.T.H. The Correlation between Daily Lens Wear Duration and Dry Eye Syndrome. *Open Access Maced. J. Med. Sci.* **2018**, *6*, 829–834. [CrossRef]
31. Mohamad Daud, S.A.B.; Binti Badarudin, N.O. Prevalence of contact lens related dry eye (clrde) among contact lens wearer in Malaysia. *Int. J. Allied Health Sci.* **2020**, *4*, 1063–1073.
32. Sapkota, K.; Martin, R.; Franco, S.; Lira, M. Common symptoms of Nepalese soft contact lens wearers: A pilot study. *J. Optom.* **2015**, *8*, 200–205. [CrossRef] [PubMed]
33. Fahmy, R.M.; Aldarwesh, A. Correlation between dry eye and refractive error in Saudi young adults using noninvasive Keratograph 4. *Indian J. Ophthalmol.* **2018**, *66*, 653–656. [CrossRef] [PubMed]

MDPI
St. Alban-Anlage 66
4052 Basel
Switzerland
www.mdpi.com

Life Editorial Office
E-mail: life@mdpi.com
www.mdpi.com/journal/life

www.ingramcontent.com/pod-product-compliance
Lightning Source LLC
Chambersburg PA
CBHW062033220926
51850CB00010B/2535
* 9 7 8 3 0 3 6 5 9 0 6 5 3 *